Applications of Supercritical Fluids in Industrial Analysis

To my wife Lynne and children Sam and Naomi
for their continued support and patience

Applications of Supercritical Fluids in Industrial Analysis

Edited by

JOHN R. DEAN

Department of Chemical and Life Sciences
University of Northumbria at Newcastle

SPRINGER-SCIENCE+BUSINESS MEDIA, B.V.

First edition 1993

Originally published by Chapman & Hall in 1993

Typeset in 10/12 pt Times by Acorn Bookwork, Salisbury, Wiltshire

ISBN 978-0-7514-0057-1

A catalogue record for this book is available from the British Library

Library of Congress Cataloging-in-Publication data

Applications of supercritical fluids in industrial analysis / edited by J.R. Dean
p. cm.
Includes bibliographical references (p.) and index.
ISBN 978-0-7514-0057-1 ISBN 978-94-011-2146-0 (eBook)
DOI 10.1007/978-94-011-2146-0
1. Supercritical fluid extraction. 2. Supercritical fluid chromatography. I. Dean, J. R.
TP156.E8A66 1993
660'.2842--dc20 92-16812
CIP

Printed on acid-free text paper, manufactured in accordance with ANSI/NISO Z39.48-1992 (Permanence of Paper).

Preface

The continued search for rapid, efficient and cost-effective means of analytical measurement has introduced supercritical fluids into the field of analytical chemistry. Two areas are common: supercritical fluid chromatography and supercritical fluid extraction. Both seek to exploit the unique properties of a gas at temperatures and pressures above the critical point. The most common supercritical fluid is carbon dioxide, employed because of its low critical temperature (31 °C), inertness, purity, non-toxicity and cheapness. Alternative supercritical fluids are also used and often in conjunction with modifiers. The combined gas-like mass transfer and liquid-like solvating characteristics have been used for improved chromatographic separation and faster sample preparation.

Supercritical fluid chromatography (SFC) is complementary to gas chromatography (GC) and high performance liquid chromatography (HPLC), providing higher efficiency than HPLC, together with the ability to analyse thermally labile and high molecular weight analytes. Both packed and open tubular columns can be employed, providing the capability to analyse a wide range of sample types. In addition, flame ionization detection can be used, thus providing 'universal' detection.

The separation of complex mixtures by chromatography provides an effective means of identifying individual components. However, as with most techniques, it is the sample preparation which is both time and labour intensive. Traditionally, Soxhlet extraction has been employed but this has several disadvantages when extracting compounds of low and medium polarity using liquid solvents. Extractions are time consuming, relatively unselective and use large amounts of hazardous solvents, so that further clean-up is usually necessary. Furthermore, after extraction and clean-up steps, the solution must usually be concentrated by evaporation prior to introduction into a chromatograph. Supercritical fluid extraction offers an elegant alternative to the currently applied extraction procedures. The ability of supercritical fluids to deposit extractants by simple depressurisation and/or cryogenic focusing means that solvent concentration can be done quickly and without heat or vacuum. These latter points are particularly important when one is seeking preservation of sensitive analytes.

It therefore seems appropriate at this stage to review the current situation regarding supercritical fluids in analytical chemistry and to consider the future possibilities to which they may be applied.

John R. Dean

Contributors

Keith Bartle	Supercritical Fluids Group, School of Chemistry, University of Leeds, Leeds LS2 9TJ, UK
Anthony A. Clifford	Supercritical Fluids Group, School of Chemistry, University of Leeds, Leeds LS2 9TJ, UK
Paul Davis	ICI plc, Pharmaceuticals Division, Safety of Medicines, Mereside, Alderley Park, Macclesfield, Cheshire SK10 4TG, UK
John R. Dean	Department of Chemical and Life Sciences, University of Northumbria at Newcastle, Ellison Building, Newcastle upon Tyne NE1 8ST, UK
Tyge Greibrokk	Department of Chemistry, University of Oslo, PO Box 1033 Blindern, 0315 Oslo 3, Norway
Steve M. Hitchen	Department of Chemical and Life Sciences, University of Northumbria at Newcastle, Ellison Building, Newcastle upon Tyne NE1 8ST, UK
Trudy K. Hoge	Dionex Corporation, Lee Scientific Division, 4426 South Century Drive, Salt Lake City, Utah 84123, USA
Václav Janda	Department of Water Technology and Environmental Engineering, Prague Institute of Chemical Technology, Technická 5, CS-16628 Prague 6, Czech Republic
Mark Kane	Department of Chemical and Life Sciences, University of Northumbria at Newcastle, Ellison Building, Newcastle upon Tyne NE1 8ST, UK
David E. Knowles	Dionex Salt Lake Technical Center, 1515 West 2200 South, Suite A, Salt Lake City, Utah 84119, USA
Tom P. Lynch	Analytical and Applied Sciences, BP Group Research and Engineering, Chertsey Road, Sunbury-on-Thames, Middlesex TW16 7LN, UK

Robert J. Ruane ICI plc, Pharmaceuticals Division, Safety of Medicines, Mereside, Alderley Park, Macclesfield, Cheshire SK10 4TG, UK

Ian D. Wilson ICI plc, Pharmaceuticals Division, Safety of Medicines, Mereside, Alderley Park, Macclesfield, Cheshire SK10 4TG, UK

Contents

1 Properties of supercritical fluids **1**

S.M. HITCHEN AND J.R. DEAN

1.1 Introduction 1
1.2 Fundamental studies of supercritical fluids 4
1.3 Choice of supercritical fluid for extraction 4
1.4 Prediction of solubility 5
References 9

2 Instrumentation for supercritical fluid chromatography **12**

T. GREIBROKK

2.1 The anatomy of a supercritical fluid chromatograph 12
2.1.1 Commercial instruments 12
2.1.2 Home-made instruments 13

2.2 The quality of the fluids 15
2.2.1 Carbon dioxide 15
2.2.2 Modifiers in carbon dioxide 15

2.3 The pumps 16
2.3.1 Pneumatic amplifier pumps 16
2.3.2 Syringe pumps 16
2.3.3 Reciprocating pumps 17

2.4 Sample introduction 17
2.4.1 Injection of solutions 19
2.4.2 Peak focusing 20
2.4.3 Direct injection 20
2.4.4 Open-split (dynamic split) injection 21
2.4.5 Timed-split injection 21
2.4.6 Sample solvent effects on peak shape and band broadening 23
2.4.7 Solvent venting under supercritical conditions 24
2.4.8 Solvent venting under reduced pressure 24
2.4.9 Solvent backflushing methods 26
2.4.10 Introduction of supercritical fluid extracts 27
2.4.11 Sample losses in the valves 28

2.5 Columns 29
2.5.1 Open tubular capillary columns 29
2.5.2 Packed columns 30

2.6 Restrictors 30

2.7 Detectors 32
2.7.1 UV and fluorescence detection 32
2.7.2 Fourier transform infrared spectroscopy (FT-IR) 33
2.7.3 Chemiluminescence detection 34
2.7.4 Flame ionization detection 35
2.7.5 Thermoionic detection (TID) or alkali flame detection 35
2.7.6 Flame photometric detection 36
2.7.7 Plasma-based detection 36
2.7.8 Electron-capture detection 37
2.7.9 Ion-mobility detection 37
2.7.10 Light-scattering detection 38
2.7.11 Mass spectrometric detection 39

References 40

3 Instrumentation for supercritical fluid extraction 46

J.R. DEAN AND M. KANE

3.1 Introduction 46

3.2 SFE instrumentation 48
3.2.1 Reciprocating pumps 48
3.2.2 Syringe pumps 49
3.2.3 Other pump modules 50
3.2.4 Sample cell 52

3.3 Depressurization systems 58
3.3.1 Restriction devices 58
3.3.2 Fixed restrictors 58
3.3.3 Variable restrictors 59

3.4 Collection of extracts 61
3.4.1 On-line SFE 61
3.4.2 Off-line SFE 63

3.5 Method development 65
3.5.1 Sample type 65

3.6 Conclusions 68

References 69

4 Supercritical fluid chromatography and extraction of pharmaceuticals **74**

I.D. WILSON, P. DAVIS AND R.J. RUANE

4.1 Introduction 74

4.2 SFC of pharmaceuticals 75
4.2.1 Alkaloids 75
4.2.2 Amphetamines 76
4.2.3 Avermectins 77
4.2.4 Barbiturates 78
4.2.5 Benzodiazepines 80
4.2.6 Beta-blockers 82
4.2.7 Cephalosporins 86
4.2.8 Cyclosporin 87
4.2.9 Erythromycin A 87
4.2.10 Mefloquine 87
4.2.11 Mitomycin C 88
4.2.12 Non-steroidal anti-inflammatory drugs (NSAIDs) 88
4.2.13 Prostaglandins 91
4.2.14 Purines and pyrimidines 92
4.2.15 Sulphonamides 93
4.2.16 Steroids 94
4.2.17 Tetracycline and oxytetracycline 97
4.2.18 Xanthines 97
4.2.19 Miscellaneous applications 98

4.3 Drugs of abuse 99

4.4 Supercritical fluid extraction 99

4.5 Conclusions 100

References 101

5 Applications of supercritical fluids in polymer analysis **104**

D.E. KNOWLES AND T.K. HOGE

5.1 Introduction 104

5.2 Supercritical fluid chromatography 106
5.2.1 Method development 106
5.2.2 Mobile phase 106
5.2.3 Detectors 107
5.2.4 Injection 108
5.2.5 Stationary phase 108

5.3 Applications of SFC for polymers 110
5.3.1 α-Olefins 110
5.3.2 Polyglycols 110
5.3.3 Ethoxylated alcohols 111
5.3.4 Polystyrenes 113
5.3.5 Polysulfides 114
5.3.6 Polysiloxanes 114
5.3.7 Polymer additives 115
5.3.8 Miscellaneous polymers 116

5.4 Supercritical fluid extraction 117
5.4.1 Off-line SFE versus on-line SFE 117
5.4.2 Method development 119
5.4.3 Extraction fluid 119
5.4.4 Extraction pressure and temperature 120
5.4.5 Extraction cell packing techniques 121

5.5 Applications of on-line extraction 121
5.5.1 Polymer additives 121
5.5.2 Hydrocarbon waxes 123
5.5.3 Polystyrene 124

5.6 Applications of off-line extraction 125
5.6.1 Polyglycols 125
5.6.2 Ethoxylated alcohols/surfactants 126

5.7 Conclusions 127

References 127

6 Applications of supercritical fluids in food science 130

J.R. DEAN

6.1 Chemical derivatization 130

6.2 Solid-phase extraction 132

6.3 Multidimensional chromatography 133

6.4 Drug and chemical residues 134
6.4.1 Dioxins 134
6.4.2 Pesticides 134
6.4.3 Insecticides 135
6.4.4 Herbicides 136
6.4.5 Mycotoxins 138

6.5 Vitamins 138

6.6 Lipids 139
6.6.1 Hydroperoxide 141
6.6.2 Cholesterol 141
6.6.3 Glycerides 142

6.7 Colours 144

6.8 Sugars 146

6.9 Waxes 146

6.10 Flavours and aromas 147
6.10.1 Caffeine 150
6.10.2 Plant products 151
6.10.3 Essential oils 153

Acknowledgements 155

References 155

7 Supercritical fluid extraction in environmental analysis 159
V. JANDA, K. BARTLE AND A.A. CLIFFORD

7.1 Introduction 159

7.2 Practice 159

7.3 Factors affecting extraction from environmental samples 162

7.4 On-line coupling of SFE to chromatographic analysis 169
7.4.1 SFE–GC 170
7.4.2 SFE–SFC 172
7.4.3 SFE–HPLC 173

7.5 Applications 173
7.5.1 Purgeable halocarbons 173
7.5.2 Chlorobenzenes and chlorophenols 174
7.5.3 Polychlorinated biphenyls 175
7.5.4 Tetrachlorodibenzo-*p*-dioxins 176
7.5.5 Organochlorine pesticides 177
7.5.6 *sym*-Triazine herbicides 177
7.5.7 Phenoxycarboxylic acid herbicides 178
7.5.8 Fuels and crude oil 178
7.5.9 Polycyclic aromatic hydrocarbons 179
7.5.10 Linear alkylbenzenesulphonates 180
7.5.11 SFE of organic compounds from adsorbent materials 180

7.5.12 Direct SFE of organics from water samples 182
7.5.13 Simultaneous SFE and derivatisation 184

References 184

8 Future developments in analytical supercritical fluid technology: some ideas from recent research 188

T.P. LYNCH

8.1 Introduction 188

8.2 Theory and understanding 188

8.3 Instrumentation 194
8.3.1 Pressure control 195
8.3.2 Pumping systems 199
8.3.3 Sampling systems 202

8.4 Special techniques 204
8.4.1 Solvents and solvent modifiers 204
8.4.2 Reactions and derivatisation 206
8.4.3 Hyphenated techniques 207
8.4.4 Reverse micelles 213

Acknowledgements 214

References 214

General index 219

Index of compounds 223

1 Properties of supercritical fluids

S.M. HITCHEN and J.R. DEAN

1.1 Introduction

A supercritical fluid is a substance with both gas- and liquid-like properties. It is gas-like in that it is a compressible fluid that fills its container, and is liquid-like in that it has comparable densities (0.1–1 g ml^{-1}) and solvating power. Supercritical behaviour only occurs when the substance is above its critical temperature and pressure. However, the converse is not always true—pressures and temperatures above the critical values do not always result in a supercritical fluid. At very high pressures (e.g. $> 10^8$ Pa) the freezing curve can rise into the supercritical fluid region and so both solid and supercritical phases can exist (Scholsky, 1989) (Figure 1.1). For analytical scale extraction the term supercritical generally refers to conditions above the critical temperature and close to the critical pressure.

Although the solvating power of liquids is generally superior to that of supercritical fluids, the use of supercritical fluids as extraction media and mobile phases in chromatography has several important potential advantages.

(1) Ease with which the solvating power of the fluid can be controlled. The solvent strength of a supercritical fluid is a function of its density, which, in turn, is a function of pressure and temperature. The relationship between pressure, temperature and density may be described by an equation of state, a number of which have been developed by various workers (Peng and Robinson, 1976; Bartle *et al.*, 1992*a*; Dohrn, 1992). The equations of state derived by Pitzer and co-workers (Pitzer, 1955; Pitzer *et al.*, 1955) have been incorporated in computer programs for the calculation of supercritical fluid parameters (SF-Solver^{TM}, Isco, Inc., PO Box 5347, 4700 Superior Street, Lincoln, Nebraska, 68505, USA). Calculation of the solvent strength of binary systems has also been described (Page *et al.*, 1991).

The effect of pressure on density for CO_2 at various temperatures is shown in Figure 1.2. The general trend (this also applies to other fluids) is for higher pressures (at a given temperature) to increase density and solvating power, while increasing temperature at a constant pressure will result in a reduction in density and hence solvent strength. These parameters (density, pressure and temperature) are therefore of prime import-

Figure 1.1 Phase diagram according to Scholsky (1989). T_p = triple point; C_p = critical point; P = pressure; T = temperature.

ance in controlling the extraction process and, although some relationships between them and solubility and rate of extraction have been developed, it is generally not yet possible to predict ideal extraction conditions on a purely physicochemical basis. However, it is often possible to predict the overall feasibility of an extraction or initial extraction condition.

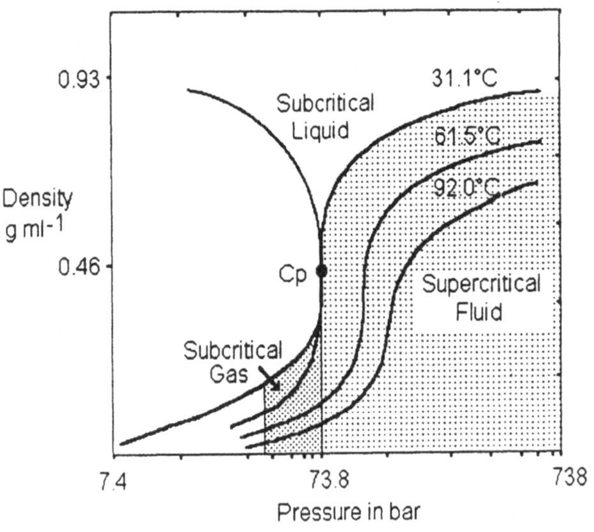

Figure 1.2 Phase diagram for carbon dioxide (C_p = critical point).

(2) High rates of mass transfer. High rates of diffusion (typically an order of magnitude higher than liquids) permit faster extraction than with liquid solvents. This is a significant property as rates of extraction are ultimately limited by the speed with which analyte molecules are transported by diffusion from the sample matrix into the bulk fluid. This process is illustrated in Figure 1.3(b). The preceding step, initial displacement of the analytes from the sample matrix, is shown in Figure 1.3(a). In chromatography the higher rates of diffusion lead to higher efficiency or the ability to have the same efficiency at higher mobile phase velocities. Diffusion coefficients are reduced as density increases, leading to less-favourable mass transfer at higher densities.

(3) Low viscosity. Supercritical fluids exhibit significantly lower viscosities than liquids (typically an order of magnitude), which provides favour-

Figure 1.3 Supercritical fluid extraction of an analyte from a sample matrix. (a) Adsorbed analyte on matrix surface in static supercritical fluid; (b) desorption of analyte from matrix surface and diffusion into bulk supercritical fluid.

able flow properties. This permits supercritical fluids to penetrate matrices with low permeability more readily than conventional solvents (although with polymer matrices swelling, caused by an appropriate solvent, will also assist access).

(4) Ease of removal from analytes. The commonly used supercritical fluids such as carbon dioxide (CO_2) and nitrous oxide (N_2O) are gaseous at room temperature and pressure and so can be separated from the analytes by decompression.

(5) Other considerations. Most substances used for analytical supercritical fluid extraction are inert, available in high purity, non-toxic and relatively inexpensive. Supercritical fluid extraction of thermally sensitive materials is possible with fluids such as CO_2 and N_2O which have low critical temperatures (31 °C and 36 °C, respectively).

1.2 Fundamental studies of supercritical fluids

Although a large number of publications have centred on fundamental studies of the interactions between supercritical fluids and solutes (Kurnik *et al.*, 1981; Johnston *et al.*, 1982; King, 1989; Bartle *et al.*, 1990*c*; Erkey and Akgerman, 1990; Bartle *et al.*, 1992*b*; Pawliszyn, 1993), lack of detailed understanding of these interactions continues. Many of these studies have involved spectroscopic investigations (Hyatt, 1984; Sigman *et al.*, 1985; Swaid *et al.*, 1985; Yonker *et al.*, 1986; Zerda *et al.*, 1986; Deye *et al.*, 1990; McNally and Bright, 1992; Sun *et al.*, 1992) based on steady-state solvatochromism to gain an insight into supercritical fluid solvent strength and fluid–solute interactions. For quantitation of solvent strength the Kamlet-Taft π^* solvent polarity scale is generally adopted (Kamlet *et al.*, 1977, 1979).

1.3 Choice of supercritical fluid for extraction

A number of compounds possess critical properties that could permit convenient application in SFE (Table 1.1). These fluids span a wide range of solvent strengths but, for practical reasons, the range generally used is more restricted (Hawthorne, 1990). Some fluids (e.g. ammonia) are usually ruled out because of corrosion or toxicity hazards. Ethane and ethene are highly flammable and their use requires special safety considerations. Methanol, although possessing a high solvent strength, is not generally employed as a supercritical fluid alone due to the rather high critical temperature; it is however commonly used with other fluids as a modifier. On the basis of solubility parameters alone, nitrous oxide (N_2O)

Table 1.1 Properties of selected fluids

Fluid	Critical temperature	Critical pressure		
	(°C)	Bar	Atm	psi
Carbon dioxide	31.1	73.8	74.8	1070.4
Ethane	32.4	48.8	49.5	707.8
Methanol	240.1	80.9	82.0	1173.4
Ammonia	132.4	113.5	115.0	1646.2
Nitrous oxide	36.6	72.4	73.4	1050.1
Xenon	16.7	58.4	59.2	847.0
Water	374.4	221.2	224.1	3208.2
Chlorodifluoromethane	96.3	49.7	50.3	720.8

would be expected to be very similar to CO_2 in terms of the range of analytes that would be expected to be readily extracted. However, notable differences in extraction efficiencies of these two fluids have been reported. Although N_2O possesses a small permanent dipole whereas CO_2 has no permanent dipole, differences in the interaction with the matrix were probably responsible, i.e. N_2O is better than CO_2 at displacing solutes from adsorption sites on the matrix. Promising results have been obtained using chlorodifluoromethane (Liu *et al.*, 1990), which has a higher dipole moment than CO_2 and can thus be expected to solubilise more polar analytes; however, few applications have been reported. The main problem with the most widely used supercritical fluid, CO_2, is the relatively low solvent strength at typical extraction pressures (80–600 atm). This limits the application of pure CO_2 to the extraction of non-polar analytes such as hydrocarbons, halogenated hydrocarbons, steroids, fats, organochlorine pesticides, etc. For more polar analytes, addition of modifiers is usually required. Again, a significant contribution to the overall effect of the addition of a polar modifier such as methanol will be in its ability to displace analyte molecules from adsorption sites on the matrix, particularly when small quantities (approximately 1%) are being used. The presence of small quantities of water should also assist this displacement process where highly polar matrices (e.g. silicacious material in soil) are involved. The type and quantity of modifier used with CO_2 in supercritical fluid extraction (SFE) is generally arrived at by trial and error as little solubility data for analytes in modified CO_2 exist; common modifiers include methanol, propanol and dichloromethane.

1.4 Prediction of solubility

The solubility of an analyte in the supercritical fluid is an important consideration when planning an extraction (Bartle *et al.*, 1991). If solubility can be predicted (Johnston *et al.*, 1989; Mitra and Wilson, 1991) then appropriate conditions for the extraction can be selected with reduced development time and effort, particularly if class-selective extractions are being contemplated. The overall feasibility of the extraction may also be estimated, i.e. if predicted analyte solubility is negligible then another fluid should be selected or the use of modifiers would be appropriate. However, prediction of solubility is complicated by the fact that often it is a range of analytes, rather than a single compound, that is to be extracted. Furthermore, analyte solubility is only one factor in determining overall extraction conditions. Other factors, such as interactions with the sample matrix, may be as, or more, important—particularly when the matrix contains adsorption sites. In the latter case, in order to remove the analyte, the supercritical fluid must be able to compete effectively for interaction with the active sites. The presence of water or other polar modifiers can prove critical in assisting this competitive process. Consideration of the role of water and other polar modifiers in normal phase liquid chromatography may be a useful model when attempting to understand the mechanisms involved here.

Some simple rules of thumb have been used to estimate solubility. The solubilising properties of supercritical carbon dioxide have been compared to hexane and benzene (see also chapter 3) but more precise methods for solubility prediction are generally required.

Limited studies have been made to correlate solute retention in chromatographic systems with solubility in supercritical fluids (Smith *et al.*, 1987; Bartle *et al.*, 1990*a*,*b*; Erkey and Akgerman, 1990). An inverse relationship between solubility in supercritical carbon dioxide and capacity factor in both supercritical fluid and reverse phase liquid chromatography was found (Bartle *et al.*, 1990*b*). However, the study was limited to five polynuclear aromatic hydrocarbons and the relationship with the reverse phase system was only followed well with naphthalene. If this relationship could be extended to other solutes it would form a useful system for prediction of solubility as much data exist in the literature on the chromatographic behaviour of compounds. This information would also probably be available in-house as chromatography is often the method of choice for the post-extraction analysis.

Most methods for the quantitative prediction of solubility have been based on correlations between solute molecular structure and solubility (Gangadhara Rao and Mukhopadhyay, 1990; King and Friedrich, 1990). Statistical mechanical models, equation of state and solution thermodynamic approaches have been investigated to correlate phase equilibria and solubility in supercritical fluids. However, the application of these methods is limited in practice as they require extensive physicochemical data and are generally based on a limited range of solutes with simple structures (e.g. aromatic hydrocarbons). Perhaps the most practical approach is that described by King and Friedrich (1990) based on solubility parameter theory.

According to King and Friedrich (1990) the understanding of solute behaviour in dense gases is assisted by consideration of four basic parameters.

(1) Miscibility pressure. This refers to the pressure (or corresponding density) at which the solute *starts to dissolve* in the fluid. The concept was introduced by Giddings *et al.* (1969) as the threshold pressure and may vary according to the technique used to monitor the solute concentration in the supercritical fluid. It may be used to choose a starting pressure for extraction.

Miscibility pressure may be obtained by solving for the interaction of the total interaction parameter, χ, and the critical interaction parameter χ_c as a function of pressure or graphically, using a plot of χ and χ_c versus pressure (Figure 1.4). Experimental confirmation of this has been noted (King, 1989).

(2) Maximum solubility pressure. This is the pressure at which the solute is maximally soluble in the supercritical fluid. Maximum solubility is predicted when the solubility parameter of the fluid equals that of the solute. The former may be estimated by application of Giddings' equation

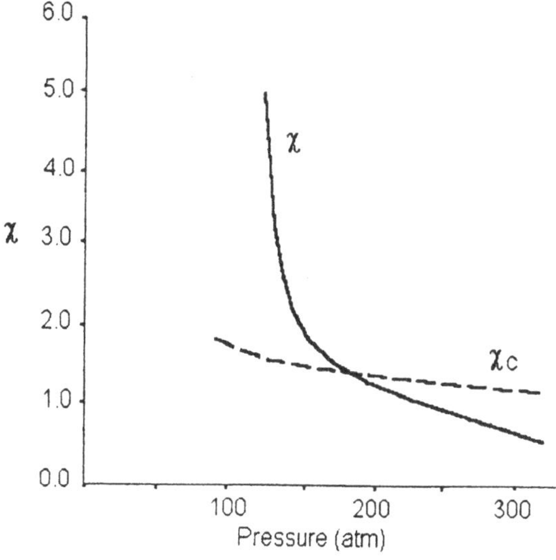

Figure 1.4 Effect of total interaction parameter, χ, on the applied pressure (χ_c = critical interaction parameter).

relating the solubility parameter of the gas to its critical and reduced state properties (Giddings *et al.*, 1969).

(3) Fractionation pressure range. This describes the range of pressures between the miscibility pressure and the maximum solubility pressure over which the solute's solubility will vary from zero to a maximum. It is in this range that enrichment of one solute over another in the supercritical fluid becomes possible by varying the pressure (or corresponding density). However, such fractionations are not generally complete.

(4) Solute physical properties. The solute's melting point is a particularly important parameter as most solutes have increased solubility in a supercritical fluid when in liquid form. The extraction temperature can therefore be significant when attempting selective extraction of solutes with widely differing melting points or vapour pressures. Temperature changes will also alter the solubility parameters.

In order to correlate molecular structure with solubility in a supercritical fluid at a particular pressure and temperature King and Friedrich (1990) used the concept of the *reduced solubility parameter*, Δ, which is defined as

$$\Delta = \delta_1/\delta_2$$

where δ_1 is the solubility parameter of the fluid and δ_2 is the solubility parameter of the solute.

The solubility parameter of the fluid, δ_1, may be calculated according to Giddings *et al.* (1969).

$$\delta_1 = 1.25\ P_c^{0.5}\ [\rho/\rho_{liq}]$$

where P_c is the critical pressure, ρ is the density of the supercritical fluid and ρ_{liq} is the density of the liquid gas.

The solubility parameter of the solute, δ_2, was calculated by King and Friedrich (1990) according to Fedors group contribution method (Fedors, 1974). Thus δ_2 can be related to molecular structure using the expression $\delta_2 = (\Delta\varepsilon/\Delta v)^{0.5}$, where $\Delta\varepsilon$ is the energy of vaporisation at a given temperature and Δv is the corresponding molar volume, which is calculated from the known values of molecular weight and density. An example of the calculation for megestrol acetate is shown in Table 1.2. Good correlation of Δ with solubility and the distribution coefficient between water and carbon dioxide was shown with a wide range of solutes.

Another method of directly relating solute structure to retention and solubility in supercritical fluids is based on molecular connectivity indices (Randic, 1975). The indices are single numbers which attempt to encapsulate how atoms are interconnected in a molecule. Using the Randic algorithm (Randic, 1975) these are readily calculated from the molecular

Table 1.2 Calculation of the solubility parameters, δ_2, for megastrol acetate according to Fedors group contribution model

CH_3 CO CH_3 $OCOCH_3$ CH_3 O CH_3

(1)

Group	$\Delta\varepsilon$(cal mol^{-1})	Δv(cm^3 mol^{-1})
5 × CH_3	5(1125)	5(33.5)
6 × CH_2	6(1180)	6(16.1)
3 × CH	3(820)	3(−1.0)
2 × HC=	2(1030)	2(13.5)
3 × C	3(350)	3(−19.0)
2 × C=	2(1030)	2(−5.5)
2 × C=O	2(4150)	2(10.8)
OCO	4300	18.0
4 ring closure 5–6 atoms	4(250)	4(16.0)
2 conjugated double bonds	2(400)	2(2.2)
	34735	328.1

$$\delta_2 = (\Delta\varepsilon/\Delta v)^{0.5} = 10.29 \text{ cal}^{0.5} \text{ cm}^{-3/2}$$

structure of the analytes. This method has only been applied to a limited range of solutes so far, but merits further investigation.

References

Bartle, K.D., Clifford, A.A. and Jafar, S.A., (1990*a*), Relationship between retention of a solid solute in liquid and supercritical fluid chromatography and its solubility in the mobile phase. *J. Chem. Soc. Faraday Trans.*, **86**(5), 855–860.

Bartle, K.D., Clifford, A.A., Jafar, S.A., Kithinji, J.P. and Shilstone, G.F., (1990*b*), Use of chromatographic retention measurements to obtain solubilities in a liquid or supercritical fluid mobile phase. *J. Chromatogr.*, **517**, 459–476.

Bartle, K.D., Clifford, A.A., Hawthorne, S.B., Langenfeld, J.J., Miller, D.J. and Robinson, R. (1990*c*). A model for dynamic extraction using a supercritical fluid. *J. Supercritical Fluids*, **3**, 143–149.

Bartle, K.D., Clifford, A.A., Jafar, S.A. and Shilstone, G.F., (1991), Solubilities of solids and liquids of low volatility in supercritical carbon dioxide. *J. Phys. Chem. Ref. Data*, **20**(4), 713–756.

Bartle, K.D., Clifford, A.A. and Shilstone, G.F., (1992*a*), Estimation of solubilities in supercritical carbon dioxide: A correlation for the Peng–Robinson interaction parameters. *J. Supercritical Fluids*, **5**, 220–225.

Bartle, K.D., Boddington, T., Clifford, A.A. and Hawthorne, S.B., (1992*b*), The effect of solubility on the kinetics of dynamic supercritical fluid extraction. *J. Supercritical Fluids*, **5**, 207–212.

Deye, J.F., Berger, T.A. and Anderson, A.G., (1990), Nile Red as a solvatochromic dye for measuring solvent strength in normal liquids and mixtures of normal liquids with supercritical and near critical fluids. *Anal. Chem.*, **62**, 615–622.

Dohrn, R., (1992), General considerations for pure-component parameters of two-parameter equations of state. *J. Supercritical Fluids*, **5**, 81–90.

Erkey, C. and Akgerman, A., (1990), Chromatography theory: Application to supercritical fluid extraction. *AIChE J.*, **36**(11), 1715–1721.

Fedors, R.F., (1974), A method for estimating both the solubility parameters and molar volumes of liquids. *Polymer Eng. Sci.*, **14**(2), 147–154.

Gangadhara Rao, V.S. and Mukhopadhyay, M., (1990), Solid solubilities in supercritical fluids from group contributions. *J. Supercritical Fluids*, **3**, 66–70.

Giddings, C.J., Myers, M.N. and King, J.W., (1969), Dense gas chromatography at pressures to 2000 atmospheres. *J. Chromatogr. Sci.*, **7**, 276–283.

Hawthorne, S.B., (1990), Analytical-scale supercritical fluid extraction. *Anal. Chem.*, **62**, 633A–642A.

Hyatt, J.A., (1984), Liquid and supercritical carbon dioxide as organic solvents. *J. Org. Chem.*, **49**, 5097–5101.

Johnston, K.P., Ziger, D.H. and Eckert, C.A., (1982), Solubilities of hydrocarbon solids in supercritical fluids. The augmented van der Waals treatment. *Ind. Eng. Chem., Fundam.*, **21**, 191–197.

Johnston, K.P., Peck, D.G. and Kim, S., (1989), Modelling supercritical mixtures: How predictive is it?, *Ind. Eng. Chem. Res.*, **28**, 1115–1125.

Kamlet, M.J., Abboud, J.L. and Taft, R.W., (1977), The solvatochromic comparison method. 6. The π^* scale of solvent polarities. *J. Am. Chem. Soc.*, **99**(18), 6027–6038.

Kamlet, M.J., Hall, T.N., Boykin, J. and Taft, R.W., (1979), Linear solvation energy relationships. 6. Additions to and correlations with the π^* scale of solvent polarities. *J. Org. Chem.*, **44**(15), 2599–2604.

King, J.W., (1989), Fundamentals and applications of supercritical fluid extraction and chromatographic science. *J. Chromatogr. Sci.*, **27**, 355–364.

King, J.W. and Friedrich, J.P., (1990), Quantitative correlations between solute molecular structure and solubility in supercritical fluids. *J. Chromatogr.*, **517**, 449–458.

Kurnik, R.T., Holla, S.J. and Reid, R.C., (1981), Solubility of solids in supercritical carbon dioxide and ethylene. *J. Chem. Eng. Data*, **26**, 47–51.

Liu, S.F.Y., Ong, C.P., Lee, M.L. and Lee, H.K., (1990), Supercritical fluid extraction and chromatography of steroids with Freon-22. *J. Chromatogr.*, **515**, 515–520.

McNally, M.E.P. and Bright, F.V., (1992), Fundamental studies and applications of supercritical fluids: a review. In *Supercritical Fluid Technology*, McNally, M.E.P. and Bright, F.V. (eds). American Chemical Society, Washington, DC, USA.

Mitra, S. and Wilson, N.K., (1991), An empirical method to predict solubility in supercritical fluids. *J. Chromatogr. Sci.*, **29**, 305–309.

Page, S.H., Goates, S.R. and Lee, M.L., (1991), Methanol/CO_2 phase behaviour in supercritical fluid chromatography and extraction. *J. Supercritical Fluids*, **4**, 109–117.

Pawliszyn, J., (1993), Kinetic model of supercritical fluid extraction. *J. Chromatogr. Sci.*, **31**, 31–37.

Peng, D.Y. and Robinson, D.B., (1976), A new two-constant equation of state. *Ind. Eng. Chem., Fundam.*, **15**(1), 59–64.

Pitzer, K.S., (1955), The volumetric and thermodynamic properties of fluids. I. Theoretical basis and virial coefficients. *J. Am. Chem. Soc.*, **77**, 3427–3433.

Pitzer, K.S., Lippmann, D.Z., Curl, R.F., Jr., Huggins, C.M. and Petersen, D.E., (1955), The volumetric and thermodynamic properties of fluids. II. Compressibility factor, vapour pressure and entropy of vaporization. *J. Am. Chem. Soc.*, **77**, 3433–3440.

Randic, M., (1975), On characterization of molecular branching. *J. Am. Chem. Soc.*, **97**(23), 6609–6615.

Scholsky, K.M., (1989), Supercritical phase transitions at very high pressures. *J. Chem. Ed.*, **66**, 989–990.

Sigman, M.E., Lindley, S.M. and Leffler, J.E., (1985), Supercritical carbon dioxide: Behaviour of π^* and β solvatochromic indicators in media of different polarities. *J. Am. Chem. Soc.*, **107**, 1471–1472.

Smith, R.D., Udseth, H.R., Wright, B.W. and Yonker, C.R., (1987), Solubilities in supercritical fluids: The application of chromatographic measurement methods. *Separation Sci. Technol.*, **22**(2/3), 1065–1086.

Sun, Y.P., Bennett, G., Johnston, K.P. and Fox, M.A., (1992), Quantitative resolution of dual fluorescence spectra in molecules forming twisted intramolecular charge-transfer states. Toward establishment of molecular probes for medium effects in supercritical fluids and mixtures. *Anal. Chem.*, **64**, 1763–1768.

Swaid, I., Nickel, D. and Schneider, G.M., (1985), NIR-spectroscopic investigations on phase behaviour of low-volatile organic substances in supercritical carbon dioxide. *Fluid Phase Equilibria*, **21**, 95–112.

Yonker, C.R., Frye, S.L., Kalkwarf, D.R. and Smith, R.D., (1986), Characterization of supercritical fluid solvents using solvatochromic shifts. *J. Phys. Chem.*, **90**, 3022–3026.

Zerda, T.W., Wiegand, B. and Jonas, J. (1986), FTIR measurements of solubilities of anthracene in supercritical CO_2. *J. Chem. Eng. Data*, **31**(3), 274–277.

2 Instrumentation for supercritical fluid chromatography

T. GREIBROKK

2.1 The anatomy of a supercritical fluid chromatograph

The essential components of a supercritical fluid chomatography (SFC) instrument consist of a mobile phase container which usually is a pressurized gas cylinder, a pump, an injector, a column in a thermostatted compartment, a restrictor and a detector. The components are similar to the components of a gas chromatography (GC) or a high performance liquid chromatography (HPLC) instrument, with the exception of the restrictor, which is needed to maintain the pressure above the critical point. Cylinders with carbon dioxide are equipped with a dip tube to allow the liquid phase to be withdrawn. With gas-phase detectors working at atmospheric pressure, the restrictor is connected prior to the detector (Figure 2.1a). If the detection is performed under supercritical conditions, the restrictor is the last component in the sequence (Figure 2.1b).

2.1.1 Commercial instruments

Commercially available complete SFC instruments have computer-controlled pumps and ovens, enabling pressure programmes, temperature programmes or density programmes to be established. A density gradient may be programmed at constant temperature or simultaneous with a temperature programme. Normally the pumps are syringe pumps, with reversible flow to enable rapid pressure equilibration between runs. The injectors are usually of the time-split type, placed in a separately heated thermostatted unit, or kept at ambient temperature, allowing nl to μl volumes to be injected and also allowing automatic samplers to be mounted. The instruments are basically constructed for the use of capillary columns, packed columns or both. A patent claim on the fundamental use of open tubular capillary columns for SFC was filed in the USA in 1982 (Novotny *et al.*, 1985) and was validated after some amendments in 1987. The flame ionization detector (FID) is often installed as the standard detector, with the restrictor mounted in the flame jet. With packed columns and modifiers the separately heated restrictor can be connected to the outlet from the high-pressure cell of an ultraviolet (UV) detector, a

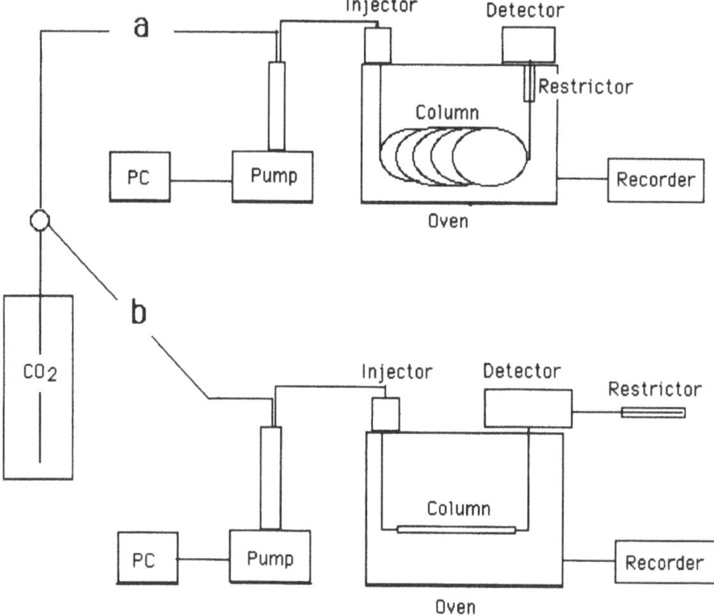

Figure 2.1 Schematic diagram of SFC instrument with gas-phase detector (a) and with the detector cell under pressure (b).

fluoresence detector or another detector compatible with the high pressures.

2.1.2 Home-made instruments

A home-made SFC instrument can be assembled with modified HPLC and GC components, or modular SFC units, depending on the level of sophistication required and on the choice of capillary columns or packed columns.

2.1.2.1 Chromatographs with open tubular capillary columns. These have column inner diameters of 50 μm and require syringe pumps to establish a stable flow. At a pressure of 350 bar, a temperature of 50 °C and a linear velocity of 2 cm s^{-1}, the volumetric flow rate of carbon dioxide is roughly 2 μl min^{-1} in the supercritical state and 1 ml min^{-1} in the gaseous state. Although a linear flow of 2 cm s^{-1} is 10 times the optimum velocity on a 50 μm column, this is a commonly used flow in order to reduce analysis time. Syringe pumps with built-in electronics for pressure programmes or with the interfacing and the software for computer control of pressure and density are available as separate units. With capillary columns, the different injection techniques utilize microvalve injectors, usually with 60–1000 nl rotors, depending on the injection mode. Large volume injections of 1 μl or more on 50 μm columns need solvent elimination techniques which in most cases have few requirements for extra equipment.

Capillary columns with different stationary phases are available from several manufacturers. Ceramic frit restrictors, integral restrictors and linear restrictors are all commercially available. The restrictor must be fitted to the detector, occasionally with an additional heating unit to prevent plugging due to the adiabatic expansion of carbon dioxide. Some detectors, such as the FID, need no extra heating in many gas chromatographs.

Instruments based on packed capillary columns and microbore packed columns need essentially the same equipment as instruments for open tubular capillary columns. Additionally, some of the reciprocating pumps are compatible with the higher flow rates of packed columns.

2.1.2.2 Instruments for packed columns. These are based on reciprocating piston pumps (or syringe pumps), standard HPLC injectors, standard-size packed HPLC columns, fluids containing modifiers and UV detection, and are relatively easy to assemble from HPLC components. In order to be able to increase the solvent strength during a run, a programme with a flow gradient, pressure gradient, density gradient, temperature gradient or modifier gradient is required. With an ordinary flow programming unit, a pressure monitor on the pump and column temperature control, tabulated densities can be used to construct density curves. Since a pressure gradient is developed over the column, long packed columns should be avoided. Some pumps are also available with a pressure control mode, but one should be aware of the fact that reciprocating piston pumps usually generate flow programmes with better stability and reproducibility than pressure programmes. With two pumps and a mixer, modifier programmes can easily be obtained.

Combinations of syringe pumps and reciprocating pumps, as well as dual syringe pumps, are not problem-free with modifier programmes, however, since the higher compressibility of carbon dioxide compared to the modifiers may cause crossflow between the pumps. The less compressible solvent is partly transferred to the pump with the higher compressible fluid. The extent of crossflow can be limited by precompression before mixing or by using a premixed fluid in each pump.

UV and fluorescence detectors require high-pressure cells, which are available from several detector manufacturers. With high flow rates and large sample throughput, oscillating-valve restrictors give few problems with plugging. The light-scattering detector is another detector compatible with high flow rates, but here the restrictor is located at the detector inlet.

2.2 The quality of the fluids

2.2.1 Carbon dioxide

CO_2 is by far the most commonly used fluid in SFC and is available in many different qualities. The analytical specifications of the gas should exclude the presence of compressor lubricants which may accidentally contaminate the whole chromatographic system. Otherwise, the choice of detector determines the requirements for the purity of the gas. Active-carbon traps and alumina traps have been used extensively to improve the quality of lower-grade carbon dioxide. The presence of hydrocarbon contaminants constitutes a considerably larger background problem with an FID than with a UV or a fluoresence detector, while halogenated impurities from seals or valves have little effect on the FID but could become of major importance with the electron-capture detector (ECD). Since oxygen and water both have electron-capturing properties, moisture and oxygen traps can be connected to reduce the ECD signal background with CO_2 gradients (Chang and Taylor, 1990). A low level of impurities is of particular importance in supercritical fluid extraction–supercritical fluid chromatography (SFE–SFC) combinations, where the impurities can be accumulated in the cold trap.

2.2.2 Modifiers in carbon dioxide

These can be purchased premixed in cylinders or can be mixed in a high-pressure container in the laboratory. The liquid phase is transferred from the container to the pump by reversing the cooled pump. The composition of a mixed liquid phase inside the container changes when the cylinder is emptied. As a consequence, one should be aware of retention differences during the lifetime of a storage container with a premixed fluid. If the last 10% of liquid is excluded from ordinary use, the variations are usually within ordinary limits of acceptance. An additional problem with syringe pumps and mixing programmes is that the remaining volume of the fluid at the start of the programme affects the compressibility factors and thus the final composition (Martin and Guiochon, 1978).

If the fluid and the modifier are pumped by reciprocating piston pumps and mixed continuously, variations of this kind are much smaller. Unfortunately, these pumps are less adaptable to the low flow rates needed with most detectors or columns in SFC. Splitting a larger flow is not easily compatible with a requirement of constant split ratio during a pressure programme.

2.3 The pumps

2.3.1 *Pneumatic amplifier pumps*

In the very early years of SFC, pneumatic amplifier pumps were utilized mainly due to their ability to create high pressures up to 2000 bar (Giddings *et al.*, 1969). However, pressure programming cannot be implemented easily, limiting the use to isobaric operations.

2.3.2 *Syringe pumps*

In order to be able to maintain a stable pulseless flow of the order of a few μl min^{-1}, under pressure control, syringe pumps have become the industry standard for capillary SFC (Figure 2.2). Varian model 8500 pumps, which were obsolete for HPLC, became attractive to SFC workers after the first report on capillary columns in SFC by Novotny *et al.* (1981). More recently new syringe pumps have become available from several manufacturers. Syringe pumps may need to be cooled during filling, in order to fill the pump completely with liquid and not partly with gas. By reducing the temperature of the carbon dioxide from ambient to approximately 15 °C, Porter *et al.* (1987) demonstrated that the pump filling increased from 60% to 90%. Cooling has another beneficial effect, to reduce the compressibility of the fluid.

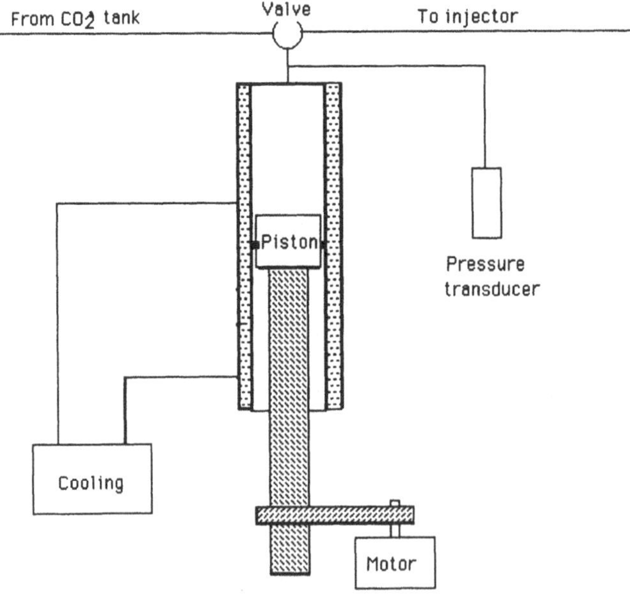

Figure 2.2 Schematic diagram of cooled syringe pump.

The most efficient way of filling the cylinder of a syringe pump is by using CO_2 cylinders with a 100 bar helium head, and many commercial cylinders are supplied with a helium head today. Particularly in laboratories with long supply lines from the cylinder to the pump, a helium head facilitates the filling. However, the introduction of helium in carbon dioxide has been shown to have adverse effects on chromatograpahic reproducibility, and other disadvantages as well (Porter *et al.*, 1987). With an efficient cooling jacket, the pump cylinder can be filled without a helium head.

During cooling operations with mixed fluids one should never forget that the miscibility of the two components is temperature-dependent.

2.2.3 Reciprocating pumps

The advantage of reciprocating HPLC pumps for SFC is the small cylinder volumes and the check valves, which largely eliminate the compressibility problems and the crossflow found with dual syringe pumps. The first HPLC pump that was modified for SFC with carbon dioxide was the diaphragm pump from Hewlett-Packard (Gere *et al.*, 1982), while the first reciprocating piston pump was the Waters model 6000 A (Greibrokk *et al.*, 1984). Cooled pump heads and cooled check valves increased the pumping efficiency and stabilized the flow of carbon dioxide (Figure 2.3). The cooling was obtained by drilling cooling channels in the pump heads or by clamp-on heat exchangers with circulating cold solvents, as with the Altex Model 110A (Mourier *et al.*, 1985*a*), the Varian 5500 (Mourier *et al.*, 1985*b*), the Perkin-Elmer Series 10 (Simpson *et al.*, 1986), the Waters Model 590 (Greibrokk *et al.*, 1987), the Gilson Model 303 (Giorgetti *et al.*, 1989), the Jeol Cap-Go 3 (Saito and Takeuchi, 1989), the Hewlett-Packard Model 1050 (Berger and Deye, 1990), the Milton Roy CP-3000 (Huang and Morgan, 1990), and the Shimadzu LC-6A (Hirata *et al.*, 1990), or by the enclosure of pump heads, valves and mixers in a box with a circulating cold solvent, as with the Jasco Familic 300 S (Sanagi and Smith, 1988). The cooling is needed to allow the check valves and the piston to work with a liquid of low compressibility. The majority of the reciprocating pumps can be expected to give stable flow at flow rates of 0.5–1 ml min^{-1} or more, while the flow rates of 0.1 ml min^{-1} or less often result in unstable flow, except with the Jeol and Shimadzu pumps, which presumably can be used at low flow rates. Today reciprocating piston pumps are used almost exclusively with packed columns and modifiers or with large volume extractors.

2.4 Sample introduction

The objective of the injector is to transfer the whole sample or a predetermined fraction of the sample quantitatively to the column in the narrowest

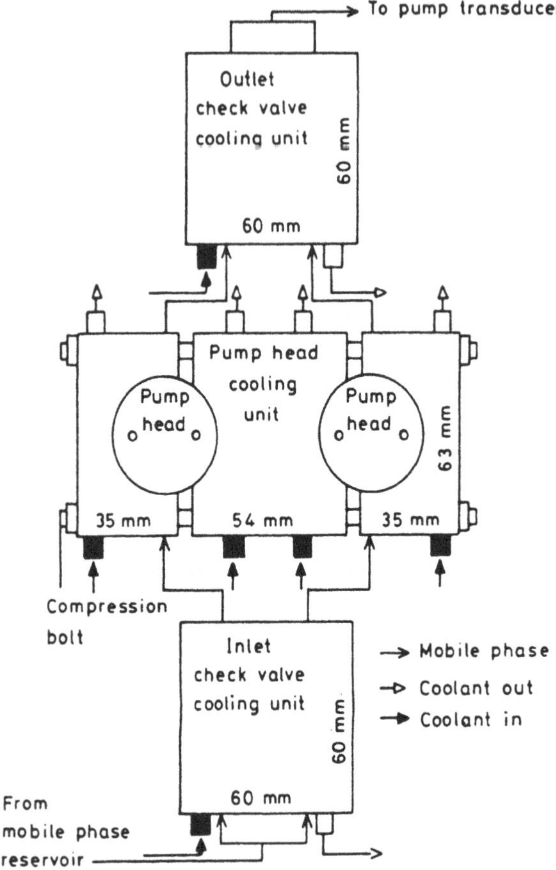

Figure 2.3 Schematic diagram of check-valve and pump-head cooling system for reciprocating pump. Reprinted from Greibrokk *et al.* (1986) with permission.

possible band. The first part of this objective, a quantitative transfer of all components, is not always obtained since sample discrimination can be found in SFC as well as in GC. The second part of the objective, transfer in a narrow band, is a process which is affected by numerous factors, such as diffusion, linear velocity, temperature, solvents, loading and mixing efficiency. Some of these factors also influence the first part of the objective, the yield of the transfer process.

Sample introduction has been given an extensive treatment in this chapter, due to the presence of several different modes of injection. Furthermore, the recent rapid development of techniques for high volume injection has not yet been implemented in commercial equipment. Also, improving a sampling method is usually easier to achieve than improving the sensitivity of a detector.

2.4.1 Injection of solutions

In most injection valves currently used in SFC, the sample is introduced as a liquid plug in the stream of fluid. With the microbore connections of capillary systems, the small internal dead volumes result in limited mixing and the major part of the sample arrives at the column inlet dissolved in a liquid composed of the solvent partially mixed with the fluid. Since most injectors are kept at room temperature or only slightly above the critical temperature of the fluid, most samples enter the column in the liquid phase. Thus, the transient elution strength of the sample may be much higher than the elution strength of the mobile phase, and this can adversely affect the peak shapes by flooding the column.

In a standard capillary system approximately half the injected amount was flushed out of a 60 nl rotor after 0.25 s (Greibrokk *et al.*, 1988). An injection time of 1 s was reported to be sufficient for quantitative transfer, since reducing the injection time from 5 s to 1 s reduced the amount entering the column by only 3%. With a 200 nl rotor, 90% was transferred to the column after 10 s (Tuominen *et al.*, 1991). The last traces of solvent, however, need time to elute from the loop. It is well known that an exponential decay injection profile is obtained by the full-injection method, because of the laminar flow of the displacing mobile phase (Scott and Simpson, 1982). By moving the valve back to the load position from the inject position by the moving injection technique (Harvey and Stearns, 1983), the solvent tail can be removed.

By injecting very small volumes by splitting or by using small sample loops, a more thorough mixing and dissolution in the mobile phase can be achieved prior to the column. Currently the nominal volume of the smallest commercially available sample loop is 60 nl. However, small-volume (<100 nl) internal sample loops are difficult to manufacture with high accuracy and the small volumes are difficult to maintain without changes, particularly with loops made from polymeric materials.

In SFC, column overload is caused rarely by the mass of the solutes, but almost always by the amount of solvent in the sample. Even 60 nl injections may prove too much for a capillary column, depending on the solvent and the retention of the solute. With a capacity factor $k' = 1$, the injected volume that produces 10% peak broadening has been calculated to be 24 nl on a 50 μm i.d. column, and with a capacity factor of 10 the volume was calculated to be 130 nl (Lee *et al.*, 1989). With an actual injection of 223 nl, 10% loss of resolution was measured for a solute with $k' = 3.2$ on a 50 μm i.d. column (Koski *et al.*, 1991).

The main disadvantage of being forced to introduce small volumes is the requirement for concentrated sample solutions. This removes the possibility of trace analysis, and the highly concentrated solutions also increase the memory effects in subsequent injections. Solute concentrations down to

5 ppm were analysed with a standard system with a 200 nl loop, direct injection and a detector with a detection limit of 0.1 ng. With split injection the limit was determined to be 150 ppm (Tuominen *et al.*, 1991). At lower concentrations solvent removal is required.

2.4.2 Peak focusing

If a 1 μl sample volume is injected on a 50 μm i.d. column, 0.5 m of the column is filled by the solvent plug alone. Consequently, peak focusing of volumes larger than this on short and narrow uncoated retention gaps is not feasible in SFC. An important function of a retention gap or precolumn in SFC is to aid in the mixing process of the solvent with the fluid, in order to reduce the elution strength of the sample band, promoting peak focusing at the column inlet. With a lowered elution strength the solutes in the front of the band slow down as they enter the column, allowing the rear part to catch up. Better mixing is, however, obtained by a short length of a wider tube and a make-up flow of fluid (Hirata *et al.*, 1990).

Precipitation of the solutes on the retention gap by evaporation of the solvent is another way of introducing the sample, but this requires a larger retention gap, inert-gas flushing and venting to the atmosphere (Davies *et al.*, 1989). By using a coated precolumn, with a smaller film thickness than on the main column, a phase ratio focusing can be obtained (Farbrot Buskhe *et al.*, 1988). With a less interacting stationary phase on the precolumn, selectivity focusing or a combination of phase ratio and selectivity focusing can be obtained. With different temperatures on the precolumn and the column, density focusing can be achieved (Davies *et al.*, 1989; Schomburg and Roeder, 1989).

2.4.3 Direct injection

Direct injection is the standard method of sample introduction in packed column SFC. In an HPLC injector, with internal or external sample loop, the sample loop is placed in the fluid path and the sample is swept into the column by the fluid. Injector temperatures exceeding room temperature are used only with solutes which need higher temperature to stay in solution, such as polymers or other high-molecular-weight compounds.

With open tubular columns and with narrow packed columns, direct injection has obvious disadvantages. If the injection is made at low density in order to obtain peak focusing at the column inlet, and the solubility of the solutes is critical, part of the sample may precipitate in the injector, resulting in poor accuracy and ghost peaks at subsequent injections. Low-density injection also results in broad solvent peaks which may last for more than 10 min, depending on the column length. Injections at higher density or higher temperature result in a narrower solvent peak, but high-density injections can have a disastrous effect on the peak shape of the peaks eluting close to the solvent (Berg and Greibrokk, 1989). Injection at high temperature should be performed with great care, since a solvent that starts to evaporate when the sample is injected in a hot injector results in sample loss and poor reproducibility.

2.4.4 Open-split (dynamic split) injection

Open-split injection has been the most widely used sample introduction method on capillary columns, at least until recently. The injection is usually performed at room temperature using a 60 nl or 200 nl sample loop in an internal loop liquid chromatography (LC) injector. In the splitter, part of the sample enters the column while the larger part of the sample moves outside the column until it is vented through a restrictor. Care must be exercised to position the column properly in the splitter assembly. Considerable peak-area variations, depending on column positions, have been reported (Richter *et al.*, 1988), and by changing the dimensions of the split assembly peak-area variations of up to two orders of magnitude were obtained (Köhler *et al.*, 1988). Both results are directly connected to the relationship between diffusion and velocity. As in gas chromatography, there is no linear relationship between peak area and split ratio, and the split ratio changes with sample viscosity and mobile-phase density. If injections are made at different densities, good quantitation is difficult to obtain.

Most split systems function better with a high split ratio, i.e. with a high proportion of the sample going to waste. A high linear velocity of the fluid passing at the outside of the column is expected to improve the homogeneity of the sample. A high linear flow in the restrictor will also reduce the probability of accumulation of precipitated solutes in the restrictor, which leads to changes in the split ratio.

Based on raw peak areas, relative standard deviations of 2–8% have been obtained by open-split injection (Köhler *et al.*, 1988). With internal standards this improved to 1% or better. With difficult solutes and certain restrictors lower performance can be expected.

2.4.5 Timed-split injection

With precise computer timing in conjunction with high-speed pneumatics to move the injection valve from load to inject and back to load, the sample is split into one portion which is placed directly on the column and another portion which remains in the sample loop. In the next loop-filling the remaining sample is removed. Less sample discrimination can be expected to be obtained with timed-split injection, compared to open-split injection.

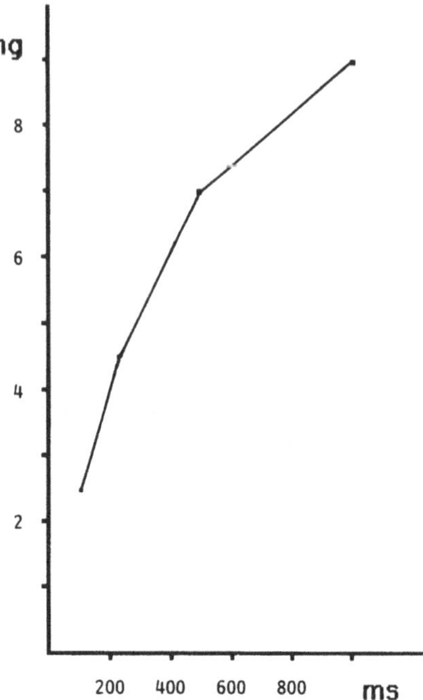

Figure 2.4 Amount of tocopherol transferred to the column as a function of the injection time for timed-split injection.

The relationship between peak area and injection time cannot be expected to be linear, and consequently nonlinear time curves are usually found (Figure 2.4). If the timed-split injector is equipped at the same time with a small open split, as on some instruments, the exact amount which is introduced on the column is not easily measurable. In general, quantitation by the internal-standards method is recommended for this injection technique, as for all methods based on splitting.

The fraction of the sample that enters the column depends on the viscosity of the sample. With solvents of higher viscosity, smaller volumes are delivered to the mobile phase during the period in which the sample loop is connected to the mobile-phase flow (Schomburg *et al.*, 1989).

Timed-split injection is usually considered to result in better reproducibility than open-split injection. With internal standards, less than 1% standard deviation was obtained with injection of *n*-alkanes (Richter *et al.*, 1988), and similar results have been obtained with injection of tocopherol (Greibrokk *et al.*, 1988). With fatty acid methyl esters, approximately the same reproducibility was obtained with timed-split as with open-split injection (Schomburg *et al.*, 1989).

A major advantage of timed-split injection is the removal of the solvent tail. The method also allows larger, more accurate, sample loops to be used with capillary columns. Timed-split injectors are currently available on most commercial instruments.

2.4.6 *Sample solvent effects on peak shape and band broadening*

Sample solvents may act as weaker solvents compared with the supercritical fluid (Hirata and Nakata, 1984), but the usual effect, particularly in supercritical carbon dioxide, is that the solvent appears to have a higher solvent strength than the fluid. The flooding or bandbroadening effect of sample solvents, which is well known from LC, is also found in SFC (Table 2.1), and split peaks are another effect of the injection of excessive volumes (Richter *et al.*, 1988; Berg and Greibrokk, 1989; Hirata *et al.*, 1989).

Although peak splitting and peak shoulders may be caused by various phenomena in the injector or at the column inlet, as also known from GC and LC, an important factor in SFC seems to be the mixing of the injection solvent with the fluid. Part of the sample elutes with the solvent plug until complete mixing is obtained on the column, while the rest of the sample is slightly retained by partitioning at the first part of the column. The splitting was found to be reduced by raising the column temperature, apparently due to increased diffusion and mixing at higher temperature. The initial pressure may also have an effect on the peak shape. At lower pressure the increased diffusion rate and the increased mixing time result in better mixing and the peak splitting disappears.

By including small mixing chambers between the injection valve and the column, the peak splitting, which otherwise resulted from 0.2–1 μl injections on 100 μm i.d. open tubular columns, could be removed (Hirata *et al.*, 1989). After further dilution of the sample solvent with make-up

Table 2.1 Flooding effect of sample solvents on column efficiency in packed column SFC. Pentadecylbenzene, with $k' = 3.8$, was injected on a 1.3 mm i.d. C_{18} column in CO_2 at 40 °C and 170 atm (data from Blilie and Greibrokk, 1985*b*, reproduced with permission)

Solvent	$N(10^3)$ 0.1 μl injected	$N(10^3)$ 1 μl injected
Acetonitrile	5.9	6.3
Methanol	—	5.9
Methyl-*t*-butyl ether	5.9	5.4
n-Octane	6.1	4.9
Dichloromethane	6.3	4.6
Tetrahydrofuran	6.0	3.7
Carbon disulphide	6.0	3.0

fluid, 4.5 μl injections were made possible in pentane and dichloromethane, but not in tetrahydrofuran (Hirata *et al.*, 1990).

2.4.7 Solvent venting under supercritical conditions

In the first solvent-venting experiments, sample volumes of up to 0.5 μl were injected on a 50 μm i.d. precolumn with complete solvent elimination (Farbrot Buskhe *et al.*, 1988). With this technique, the solvent is more or less completely separated from the solutes on the precolumn by a partitioning mechanism, and the solutes are transferred to the main column and focused at the column inlet (Figure 2.5). The major part of the solvent is vented through a restrictor at a rapid (10–15 cm s^{-1}) but controlled flow rate. With a 2 m precolumn the venting time is only 15–20 s. Too long venting times discriminate the early-eluting peaks and too high starting pressure degrades the peak shape (Berg and Greibrokk, 1989). The peak focusing at the column inlet is based on phase ratios, by using a thicker film on the column than on the precolumn, or by partition coefficients, by using a more retaining stationary phase on the column, or both. A sample recovery of 100% was obtained with palmitic acid, cholesterol, tripalmitine and cholesteryl palmitate. With a 7 m × 50 μm i.d. precolumn, injections of 1 μl samples were performed on 50 μm i.d. columns, without discrimination (Berg *et al.*, 1992).

Based on absolute area measurements, the solvent-venting technique resulted in better injection reproducibility (RSD = 2–5%), compared to timed-split injection (RSD = 6%) and open-split injection (RSD = 8%) on the same instrument. Addition of an internal standard reduced the relative standard deviation to less than 1%. With a minimum detectable quantity of 0.02–0.1 ng, sample concentrations of 0.1–0.5 ppm are within reach.

This method of solvent elimination, which is based on a rough separation of solvent and solutes by partitioning on a precolumn, seems likely to be a method for introduction of sample volumes not larger than 1–2 μl on 50 μm i.d. columns. The valve between the precolumn and the column produces some band broadening, which must be reversed by peak focusing at the column inlet. The valve gives, however, additional opportunities, e.g. to use the system for coupled-column separations.

2.4.8 Solvent venting under reduced pressure

Solvent venting by pressure reduction after mixing and dilution with supercritical CO_2 (Hirata *et al.*, 1991) or by purging the sample with a gas (Lee *et al.*, 1989) are techniques which also require a precolumn or a retention gap. By purging the sample into the precolumn with an inert gas, e.g. by GC conditions, the solvent is evaporated in the gas flow and carried rapidly through the column and out of the vent valve (Figure 2.5). After

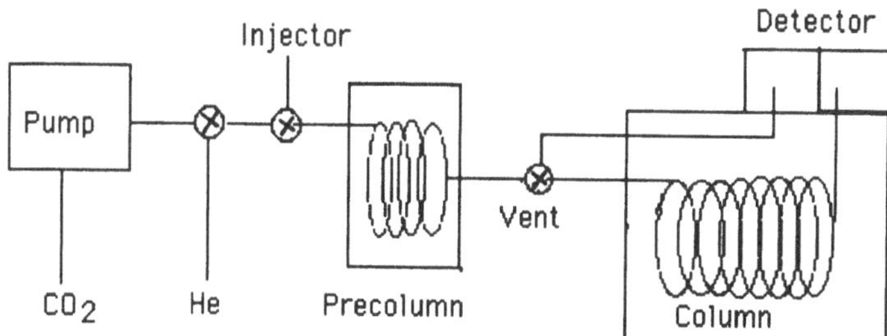

Figure 2.5 Instrumentation for two injection modes: solvent venting at supercritical conditions or solvent elimination by purging with an inert gas (He).

the solvent is evaporated and the solutes are coated on the walls of the precolumn, the purging valve is switched back to the carbon dioxide line, and liquid or supercritical CO_2 is introduced. The solutes that dissolve in the flow are focused by a temperature gradient at the column entrance (Lee *et al.*, 1989) or a temperature gradient on the precolumn (Liu *et al.*, 1991). The temperature gradient on the precolumn can be obtained by passing an electric current through a conductive paint coated on the outer surface. Rapid expansion of carbon dioxide in a wide precolumn (20 cm × 200 µm i.d.) and a temperature difference between the precolumn and the column added to the band focusing (Liu *et al.*, 1991). One advantage with gas-purging injection is the opportunity for multiple injections. For strongly retained solutes, *n*-alkanes with 32–38 carbon atoms, five multiple injections of 1 µl resulted in complete recovery. The reproducibility of 0.5 µl injections was RSD = 4–13%, based on raw areas. Compared with split injection the column efficiency was reduced by 15–20%. By using longer narrow precolumns with 11–15 µm i.d., the efficiency improved to the level of split injection.

Another gas-purging method, with helium, made use of a 1.5 m × 100 µm i.d. uncoated retention gap (Berg *et al.*, 1992). By injecting 1 µl samples of glycerides, triglycerides were completely recovered, while diglycerides suffered a 30% loss. Even after 10–50 multiple injections of 1 µl each, the triglycerides were completely recovered. A 10 µl injection of chlorinated fatty acids in chloroform is shown in Figure 2.6. The reproducibility of the 1 µl injections was RSD = 6%, based on raw areas.

With a combination of sample dilution and solvent venting, samples of up to 100 µl of triglycerides were injected on a 50 µm i.d. precolumn in combination with a 100 µm i.d. column (Hirata *et al.*, 1991). Concentrations as low as a few ppb could be analysed when solvents of very high purity were utilized.

The major efforts towards solvent elimination have been exhibited in connection with the use of open tubular capillary columns, but Dean and Poole (1989) also made a study of similar techniques for packed microbore

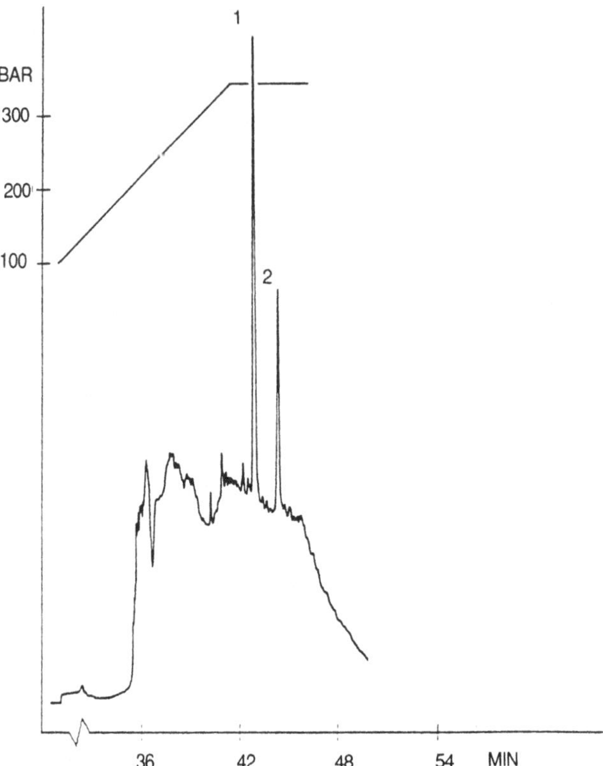

Figure 2.6 High-volume injection (10 μl) of chlorinated fatty acids in chloroform on a retention gap, 1.5 m × 100 μm i.d., connected to a 10 m × 50 μm i.d. SB-Biphenyl-30 column, by the multiple-injection helium-purging technique. Peak identification: (1) 9,10-dichlorostearic acid; (2) 9,10,12,13-tetrachlorostearic acid. Reprinted from Berg *et al.* (1992) with permission.

(1 mm i.d.) columns. By gas-purging the sample into a steel retention gap, 1–10 μl injections of *n*-alkanes in 0.1–1 ppm concentrations resulted in high peaks with good peak shapes. The reproducibility of 1 μl injections of 1–10 ppm concentrations was 0.7%, by the internal-standards method.

Packed precolumns can also be used for solvent elimination. A 6 cm × 200 μm i.d. capillary, filled with polyoctylsiloxane particles, was connected to a 50 μm i.d. column. With a valve which kept high pressure on the main column, together with the application of gas purging and backflush elution from the packed column, the column efficiency was essentially equal to the efficiency obtained by split injection (Koski *et al.*, 1991).

2.4.9 *Solvent backflushing methods*

In order to try to avoid precolumns and valves a method for eliminating the sample solvent without long solvent elution times was developed (Lee *et al.*, 1989). The instrumentation incorporates a vent line with an on/off valve. The sample is injected with a split vent valve closed. After a time delay of approximately 1 min, the injection valve is closed and the split valve is opened simultaneously with a rapid negative pressure ramp. The rapid depressurization causes a density gradient to propagate along the column from the inlet, and a reversed flow develops in the first part of the column. The solvent backflushes through the open vent line and the solutes coat the column wall. By reversing to a positive density programme, the chromatogram is developed. A similar method, but without the negative pressure ramping, was called delayed split injection. Preliminary measurements from the injection of 0.2 μl volumes with the solvent backflush technique showed a recovery of 68% of naphthalene and 85% of coronene.

Another backflushing method, called post-injection solvent venting (Hawthorne and Miller, 1989), making use of a short (7 cm) wide (0.3 mm) retention gap and syringe injection, allowed injection of 0.5 μl volumes on the depressurized column, reportedly with detection limits of 0.2 ppm of alkanes. Negative pressure ramping was not included, possibly due to the wide injection port and retention gap which facilitates rapid depressurization. The backflushing started 15 s after the pressure program was initiated and lasted for 5 s. The peak area reproducibility was approximately 10% on raw areas and 0.9–2.5% with an internal standard.

In conclusion, sample volumes of approximatley 1 μl can be injected on 50 μm i.d. columns with current solvent elimination methods. If there is a need to inject considerably larger volumes, the choice is between multiple injections with gas purging, combined dilution and venting, and sample concentration on packed precolumns. Whether the sample is injected in multiple injections or in a continuous process with a make-up fluid is a matter of taste, since the effect is largely the same. Whether packed precolumns can be utilized is completely dependent on the properties of the sample.

2.4.10 Introduction of supercritical fluid extracts

The method for transferring an aliquot of the extract or the complete extract depends on the extraction procedure. In dynamic extractions the fluid is pumped continuously through the extractor to the collector. If a static extraction is performed, an aliquot is sampled by a valve at intervals. In analytical procedures dynamic extraction requires the presence of a sample concentrating unit, based on cooling (cold trap), heating (density focusing) in combination with a precolumn, or solute precipitation in a piece of tubing by venting to the atmosphere. Depending mostly on the volatility of the sample constituents, the solute collection will be more or less quantitative.

An advantage of using the mobile phase for extraction is that the whole extract can be expected to elute. As for other chromatographic systems, there is never a guarantee that all the constituents of a sample dissolved in a solvent will be eluted, depending on the solubility in the mobile phase. However, one should not forget that if the extraction has been performed at high density, elution of all components can be expected to require a density gradient with a high-density end. The only possibility of losing part of the sample then is by using columns with higher adsorptive properties than the original sample matrix.

2.4.11 Sample losses in the valves

With pressures possibly approaching 500 bar, occasional leaks are bound to occur in a rotating valve. Many valves are guaranteed leak-proof to 400 bar, but even below this limit leaks arise. Leaks are often a result of the impact of solid particles between the rotor and the stator, particles coming from the sample, from the fluid container or from the end of the connecting tubing. Particle filters should be inserted in the line prior to the injector. Fused silica tubing, at both the inlet and the outlet, should be connected with care to avoid splintered ends. Samples with suspended particles need to be filtered.

Leaks are, however, not so easily detected since the gas flow which may escape is very low. The reduced peak height, which is the result of a leak in the valve, could also have other causes, such as detector problems. A close inspection of the stator and the rotor will usually reveal whether there is a leak.

By injecting the sample dissolved in a volatile solvent in a hot injector, part of the sample will be lost. Even when the injector is not intentionally heated, waste oven heat may warm the valve to a temperature where the expanding gas from the evaporating diluent can expel part of the liquid sample, depending on the valve port assignment. A waste-port restrictor or a valve cooling system can solve this problem (Chester and Innis, 1989).

If the yield of the transfer to the column is unequal for the solutes, the term discrimination is commonly used. Discrimination may take place when a part of the sample is split away, as with open-split injection, solvent elimination and similar methods. Volatile or rapidly eluted components are transferred to the column to a smaller extent than less volatile, more retained components. Claims of no discrimination with split systems should always be regarded as sample-related statements, and the actual extent of discrimination in every case needs to be determined with standard solutions. For the majority of applications in SFC, however, discrimination effects are not of major significance.

2.5 Columns

Open tubular capillary columns are most useful for separations requiring high-efficiency separations, for complex samples and as the analytical column in LC–SFC, SFC–SFC and SFE–SFC combinations. The columns are compatible with virtually all detectors in SFC, but the main applications are with solutes not requiring fluids containing organic modifiers.

Packed columns are most useful for high-speed separations requiring a moderate column efficiency, for samples containing fewer components needing to be separated or for group separations. With fluids containing organic modifiers, the UV detector, the fluorescence detector and the light-scattering detector are easily compatible with packed columns. Some gas-phase detectors are easily interfaced too, particularly with narrow bore columns.

2.5.1 Open tubular capillary columns

The standard inner diameter of the capillary columns is currently 50 μm. This is a compromise between the wish for higher efficiency with narrower columns, the practical problems of coating the stationary phase on very narrow columns and the requirement for column loadability. The optimum linear velocity is approximately 0.2 $cm\,s^{-1}$, depending on the density (Fields and Lee, 1985), but for most practical purposes ten times higher velocities are used. At low density, 10 000 plates m^{-1} can be obtained, while high densities (0.75 $g\,m^{-1}$) may give only 2000 plates m^{-1}. Accordingly, with density gradients the first peaks are expected to elute with higher efficiency than the later peaks. Most applications utilize 5–10 m long columns.

The immobilized stationary phases generally contain polysiloxanes as the polymeric backbone. With the methyl and *n*-octyl phases, dispersion interactions are responsible for the retention mechanism. With the 30% biphenyl phase, which is usually preferred over the 50% phenyl phase, polar solutes are better retained due to induced dipolar interactions. Polar stationary phases such as the cyanopropyl phase have excellent compatibility for solutes with carboxylic groups. In addition to the polysiloxanes of different polarity and the polar poly(ethylene glycols), liquid-crystalline phases and chiral phases exist which separate according to three-dimensional principles.

A large selection of columns coated with immobilized stationary phases with a film thickness of 0.1–0.5 μm is currently commercially available.

2.5.2 *Packed columns*

With inner diameters of 0.2–5 mm, packed analytical columns result in fluid flow rates of 0.05–10 ml min^{-1} or in gas flows of approximately 10–1000 ml min^{-1}. Clearly not all detectors are compatible with all sizes of packed columns. As an example, FIDs may not be able to accommodate columns of >1 mm i.d., resulting in approximately 100 ml min^{-1} of gas, depending on the fluid density. The columns are currently 5–20 cm long, packed with 3–5 μm particles, having an optimum linear velocity of 1–3 cm s^{-1} with 5 μm particles, depending on the density (Schoenmakers and Uunk, 1987). Since most applications on packed columns start at a velocity lower than optimum, the part of the efficiency term based on the mobile-phase velocity may actually improve with the higher flow obtained at higher density, in contrast to the situation on open tubular columns. However, the pressure drop over the column, which adversely affects the peak width of well retained compounds, counteracts the first effect.

Many HPLC columns have found use in SFC, but the problem of active sites on silica-based packings can be even more serious in SFC than in HPLC (Blilie and Greibrokk, 1985*a*; Levy and Ritchey, 1986; Yonker and Smith, 1986). Recently deactivation procedures resulting in polymer-encapsulated phases have led to improved peak shapes with polar analytes (Ashraf-Khorassani *et al.*, 1989; Payne *et al.*, 1990). New polymeric stationary phases have been developed for SFC (Yang, 1989), and the properties of polymeric packings have been studied (Gemmel *et al.*, 1989; de Weerdt *et al.*, 1990). Effects of elevated temperatures on packed columns were examined by Saunders and Taylor (1989).

With packed narrow-bore fused-silica columns, the design of the column ends is extremely important, even more than in HPLC. An overview of packed capillary-column technology was presented by Hirata (1990).

2.6 Restrictors

Most SFC instruments utilize fixed restrictors, particularly with equipment designed for narrow-bore columns. With a fixed restrictor the mass flow of fluid changes significantly during a density programme. This affects the linear velocity and the column efficiency, and also the compatibility with some detectors. With variable restrictors, density programmes can be run at constant mass flow. Although many attempts have been made to construct variable restrictors, it is an extremely difficult task to accommodate them with narrow-bore columns and gas-phase detectors without complicated constructions that usually result in significant band broadening, and no real breakthrough has been made so far.

Variable restrictors are, however, commonly used with UV and fluorescence detectors, where the detector is prior to the restrictor. Back-pressure regulation via heated valves appears to be the most prevalent method, as shown by Gere *et al.* (1982). More recent developments for mass flow control include an oscillating valve (Saito *et al.*, 1988) and a split effluent system (Janssen *et al.*, 1990). Oscillating valves are currently used extensively with extraction units, due to the absence of plugging.

Several fixed restrictor designs have been investigated over the years (Figure 2.7). The linear restrictor (Fjeldsted *et al.*, 1983*a*), which is simply a piece of fused-silica tube with i.d. 5–10 μm, is still used for narrow columns, but now mainly with larger diameters for extraction purposes. Non-volatile compounds may precipitate in this restrictor, due to the long pressure-reduction zone. The robot-drawn fused-silica tapered restrictor (Chester *et al.*, 1985) represented a significant improvement, but the hair-thin tip with a 1–2 μm orifice breaks easily. A more robust construction was the integral restrictor (Guthrie and Schwartz, 1986), made by polishing the end of a fused-silica tube carefully closed by heat.

The ceramic frit restrictor, made according to the procedure of Cortes *et al.* (1987), was another significant improvement. The restrictor, which contains a porous frit with multiple fluid pathways, is not easily plugged and is currently the most widely used restrictor due to this fact.

Metal (platinum, platinum–iridium or steel) restrictors are made by pinching the end of a piece of tubing. This is a simple method of creating a restriction, which works reasonably well except at low flow rates. The advantage of metal restrictors is the good heat-conducting properties, but pinched restrictors become plugged quite frequently. In order to avoid plugging of the restrictor with solid carbon dioxide resulting from the cooling caused by expansion from the supercritical to the gaseous state or

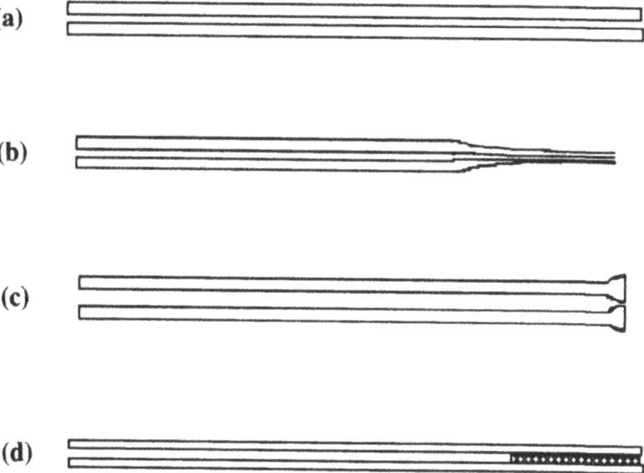

Figure 2.7 Schematic cross-sections of restrictor configurations: (a) linear restrictor; (b) tapered design; (c) integral restrictor; (d) ceramic frit restrictor.

by solutes of low solubility, the restrictor must be heated or the fluid must be heated prior to the restrictor. The heat supplied must, as a minimum, be sufficient to keep the fluid from freezing, but fluid temperatures of at least 80–100 °C are needed to obtain good flow characteristics. With detectors like the FID, the detector temperature is usually set to approximately 350 °C. The actual temperature of the fluid is not known, but much lower. With detectors where the temperature affects important properties of the aerosol coming from the restrictor, such as the light-scattering detector, heating of the restrictor must be fine-tuned. For the majority of detectors this is not the case.

2.7 Detectors

SFC incorporates elements from GC as well as from HPLC, including the detectors. With capillary columns and carbon dioxide practically all GC detectors and many HPLC detectors can be utilized. With packed columns and organic modifiers, the number of detectors available is more limited.

2.7.1 UV and fluorescence detection

The UV detector responds only to compounds with UV-absorbing functions (chromophores). With photodiode-array detectors, on-the-fly spectra can be obtained to assist in compound identification, as shown for packed column SFC (Jinno *et al.*, 1986) and for capillary columns (France and Voorhees, 1988). The detection is performed under supercritical or subcritical conditions with the restrictor connected to the detector outlet. With standard-size analytical columns. HPLC detectors with pressure-resistant flow cells are utilized (Gere *et al.*, 1982). Detection in the liquid state, by lowering the temperature, is in principle preferred to detection in the supercritical state, since supercritical fluids are more sensitive to refractive-index changes as a function of density, compared to liquids.

With narrow-bore columns, the detection is performed either on a fused-silica tube with large diameter than the column, or directly on the column with the polyimide coating removed (Kuei *et al.*, 1987; Fields *et al.*, 1988). With 50 μm i.d. columns, the maximum permitted cell volume to avoid loss of resolution has been calculated to be 50 nl (Peaden and Lee, 1983). Capillary cuvettes with diameters of 0.2–0.3 mm allow strongly absorbing compounds like phenanthrene and chrysene to be determined at the low-ng level.

The short path lengths obtained with on-column techniques reduce the detection limits. With a Z-shaped fused-silica flow cell, a 2 cm path length was obtained, increasing the UV response, however, partially at the expense of the noise level (Chervet *et al.*, 1989). Good chromatographic performance was obtained.

Another direction for absorbance-based detection in the future may be nonconventional absorbance, such as the thermal lens principle, without the path-length dependence of conventional methods (Leach and Harris, 1984). With plain CO_2 as the mobile phase, the UV detector can be used down to 200 nm. Some baseline drift can be noticed, mainly from refractive-index changes (Fields *et al.*, 1988). With alcohols as modifiers, no baseline drift was observed at 254 nm (Blilie and Greibrokk, 1985*b*).

Fluorescence detection can be performed the same way as absorbance detection, either with conventional pressure-resistant flow cells in HPLC detectors or with on-column methods (Fjeldsted *et al.*, 1983*b*; Fjeldsted and Lee, 1984). Laser-induced fluorescence in supersonic jet spectroscopy is a recently developed tehnique with a potential still not determined (Simons *et al.*, 1989).

2.7.2 Fourier transform infrared spectroscopy (FT-IR)

Absorption of energy in the infrared region with a chromatographic system connected on-line requires faster and more sensitive detectors than dispersive infrared spectrophotometers. In FT-IR, infrared radiation which has been modulated is passed through the flow cell or another sample collection unit and is detected by a mercury–cadmium telluride detector. Plots of infrared intensity versus time (interferograms) are stored in a computer. The computer performs a Fourier transform upon the interferograms. Compared to the background, a transmittance or absorbance IR spectrum is produced, as a function of wavenumbers. Rapid spectral acquisition and averaging are possible since all the resolved elements of the IR spectrum are collected simultaneously.

2.7.2.1 On-line detection. With a high-pressure flow cell, real-time monitoring of the effluent is possible. The cell is a cylinder with windows of calcium fluoride, zinc selenide or zinc sulphide and the restrictor is connected to the outlet (Shafer and Griffiths, 1983; Olesik *et al.*, 1984; Hughes and Fasching, 1985; Johnson *et al.*, 1985; French and Novotny, 1986; Morin *et al.*, 1986; Wieboldt *et al.*, 1988; Raynor *et al.*, 1988; Shah and Taylor, 1989). The advantage of an on-line system is that chromatograms based on the total IR absorbance can be reconstructed, using the Gram–Schmidt algorithm. The chromatograms can be supplied with structural information from the IR spectra of each component. The disadvantages are that a limited number of spectra can be accumulated and that few fluids (xenon is an exception) are IR-transparent. Thus, IR spectra in carbon dioxide are not directly comparable to the library spectra obtained without interfering fluids. Unfortunately xenon is very expensive (French and

Novotny, 1986). With a stopped-flow technique, Raynor *et al.* (1990) demonstrated that the quality of the spectra could be improved.

2.7.2.2 With solvent elimination. By depositing each component on a moving window of KBr or ZnSe (Shafer *et al.*, 1984; Fuoco *et al.*, 1989), diffuse reflectance spectra or transmission spectra of the components can be measured without solvent interference. The sample deposition is obtained with the retrictor positioned 50 μm above the window (Figure 2.8), and the window is transferred to the spectrometer either continuously or in an off-line mode. Most measurements have been performed in the off-line mode, with the window mounted under the FT-IR microscope. Volatile components will be lost with the solvent. Smaller amounts of sample can be detected than with the on-line methods, well into the picogram range, depending on the compound.

2.7.3 Chemiluminescence detection

Some molecules can be transferred to an excited state not only by light, but also by chemical reactions, and the amount of light emitted while returning to the ground state (chemiluminescence) can be measured.

The sulphur chemiluminescence detector (SCD) is based on the reaction of sulphur monoxide with ozone, emitting light at 320–390 nm (Foreman *et al.*, 1988). The sulphur monoxide is produced by combustion of sulphur-containing components in a hydrogen–oxygen reducing flame. By connecting the SCD to the outlet from an FID, amounts as low as 35 pg of sulphur were detected with carbon dioxide as mobile phase (Chang and Taylor, 1990). After flame gas optimization, the minimum detectable quantity of aromatic sulphur compounds was determined to 10–1400 pg (Pekay and Olesik, 1990).

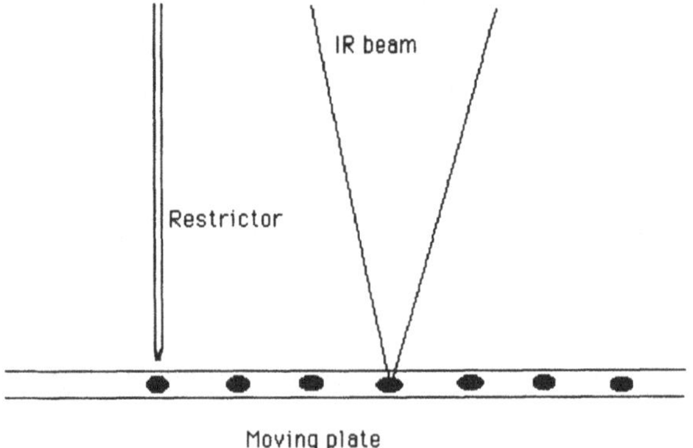

Figure 2.8 Fourier-transform infrared spectroscopy coupled off-line with SFC.

Capillary columns give the best compatibility with the SCD since the higher flow rates of packed columns can lead to signal quenching.

2.7.4 Flame ionization detection

The FID has become the most frequently used detector in SFC, as first shown by Sie *et al.* (1966), but mainly after the reports by Fjeldsted *et al.* (1983*a*) and Chester (1984) with capillary columns and Rawdon (1984) with packed columns. For a literature review see Richter *et al.* (1989). Standard FIDs can be used with no or small modifications, with the restrictor positioned in the flame jet or a few mm below the opening. With the flow rates of capillary columns and narrow packed columns, comparable sensitivity to GC has been reported after adjusting the flame gas conditions, with carbon dioxide as the mobile phase (Pekay and Olesik, 1989).

Organic modifiers to the mobile phase, with the exception of formic acid, give a background which is too high to be tolerated. Less than 1% of formic acid is soluble in carbon dioxide (Crow and Foley, 1991).

Other mobile phases, such as nitrous oxide and sulphur hexafluoride, are also compatible with the FID. A higher background is obtained with nitrous oxide than with carbon dioxide (Schaefer, 1972; Baastoe and Lundanes, 1991). The use of SF_6 requires a gold-plated collector to avoid corrosion by HF (Schwartz and Brownlee, 1986; Fields and Grolimund, 1989).

2.7.5 Thermoionic detection (TID) or alkali flame detection

The effluent is ionized on a rubidium or caesium bead which is heated electrically (to 400–800 °C) or in a hydrogen–oxygen flame. The alkali metal stimulates the formation of ions from electronegative species, particularly nitrogen and phosphorus, compared to carbon compounds (Farwell *et al.*, 1981). The first study with SFC in carbon dioxide and in nitrous oxide was performed by Fjeldsted *et al.* (1983*a*). Different modes of ionization were demonstrated by West and Lee (1986), recent developments in detector design were shown by David and Novotny (1989), and variables of the detector were optimized by Mathiasson *et al.* (1989). Detection limits were reported as 120 fg P s^{-1} and 2 pg N s^{-1} by David and Novotny (1989) and a minimum detectable quantity of 50 pg–2 ng, depending on the mode, was reported by West and Lee (1986). In the analysis of amines, a sensitivity comparable to GC was reported by Mathiasson *et al.* (1989). Methanol has been added to nitrous oxide (Mathiasson *et al.*, 1989) which tolerates less than 0.8% methanol, and to carbon dioxide (Greibrokk *et al.*, 1987) in mixtures of up to 7%. Unfortunately, the lifetime of the bead is shortened with polar modifiers in the fluid.

Heating the bead in a flame was found in early work to result in low sensitivity in SFC. Recently, however, Mol *et al.* (1991) showed that by adjusting the hydrogen flow, the air flow and the bead position, very promising results were obtained in phosphorus-selective determination of pesticides. With packed capillary columns and modifiers added to carbon dioxide, the minimum detectable quantity was 15–65 pg of pesticides. A linearity of four orders of magnitude was obtained, with an optimal sensitivity of 55–128 fg P s^{-1}.

2.7.6 *Flame photometric detection*

In the flame photometric detector (FPD), a low-temperature plasma produces atoms and simple molecular species. Some of the species are excited by collision with flame gas constituents, and the emitted light while returning to the ground state is measured. The detector is used for phosphorus, at 580 nm, and for sulphur, usually at 365 nm. The phosphorus line is determined as the emission by excited HPO and the sulphur as emission by excited S_2. Markides *et al.* (1986) determined the minimum detectable quantity as 0.5 ng of parathion in the phosphorus mode and 25 ng of benzothiophen in the sulphur mode. Later, Pekay and Olesik (1989) studied the flame gas chemistry in SFC and the optimizations that could be obtained. The detection limits improved and were close to reported values for GC. With a sharply tapered restrictor, the background was independent of the pressure, in contrast to earlier findings with other restrictors.

2.7.7 *Plasma-based detection*

The excitation of atomized elements can be increased by an inductively coupled plasma (ICP), by a microwave-induced plasma (MIP) or by a radio-frequency plasma (RFP). The temperature of the ICP may be as high as 10 000 °C, while the temperature of the MIP is lower, 1000–3000 °C. The addition of carbon dioxide in general results in an increased background and a lowered signal (Fujimoto *et al.*, 1987, 1989). The ICP provides the best detection for metallic elements, with a minimum detectable quantity of 60 pg of Fe (Olesik and Olesik, 1987; Jinno *et al.*,1990). With its high purchase cost and high consumption of argon, the ICP is an expensive detector. Even more expensive, but with better sensitivity, is the SFC–ICP–MS (Shen *et al.*, 1991). With the mass spectrometric analyser, 0.03 pg of tetrabutyl-tin could be determined.

In MIP detectors, the plasma is sustained in a surface wave device called a surfatron (Luffer *et al.*, 1988) or in a resonant cavity (Motley *et al.*, 1989). Increasing amounts of carbon dioxide had a dramatic effect on the detection limit (Rivière *et al.*, 1988). By using a near-infrared emission line for chlorine, the detection limit for Cl was determined to 40 pg s^{-1} (Zhang *et al.*, 1991). Boronate esters of hydroxyl compounds have been determined with a sensitivity of 25 pg s^{-1} of boron (Luffer and Novotny, 1991).

The RFP is sustained by a discharge between two electrodes at a frequency of 330 kHz (Skelton *et al.*, 1988). By using the near-infrared region spectra, molecular interferences are reduced as well as background from CO and CO_2 emissions. The baseline was, however, affected by the density of the carbon dioxide mobile phase. The detection limit of sulphur was determined to be between 25 and 180 pg s^{-1}.

The results so far should be judged by the fact that element-specific detection with plasma-based detectors is still in its early stage in SFC.

2.7.8 *Electron-capture detection*

The ECD, which is one of the most sensitive of the selective GC detectors, has been utilized for SFC without modifications (Kennedy and Wall, 1988; Chang and Taylor, 1990). Halogenated compounds were determined in amounts below 1 pg. Solutes of low volatility may, however, require a heated restrictor interface to increase the rate of vaporization.

Highly purified gases were required. SFC-grade carbon dioxide, which contains <3 ppm water and <5 ppm oxygen, was purified with oxygen and moisture traps, since both water and oxygen have electron-capturing properties. The reaction gases (10% methane in argon) were of high purity too. With high ratios of CO_2 to reaction gas, a nonlinear response was obtained. A reaction gas flow of 20–30 ml min^{-1} was recommended to maintain linearity and improve sensitivity.

Since carbon dioxide is a weak electron-capturing gas that can act as a scavenger for thermal electrons produced in the detector, a high flow rate will reduce the standing current and thereby the dynamic range. Even so, the detector has been shown to be compatible with the flow rates of packed capillary columns with an i.d. of 0.2 mm (Moulder *et al.*, 1991).

2.7.9 *Ion-mobility detection*

The ion-mobility detector can be viewed as an ECD with an ion-separation chamber (Baim and Hill, 1982). In contrast to the ECD, water and oxygen in the fluid are not only allowed but added, since secondary reactant ions like $(H_2O)_nH^+$ are formed. The reactant ions are used to ionize neutral solutes. Reactant ions and product ions are repelled into the entrance gate by a strong electrical field, positive or negative, at the top (Figure 2.9). The detector can be used in a total ion mode, for positive or negative ions, or in a selected ion mode. In the selected ion mode, ions of selected mobilities are permitted to pass through the separation region and are detected at the collector by keeping the exit gate open for a predetermined delay time

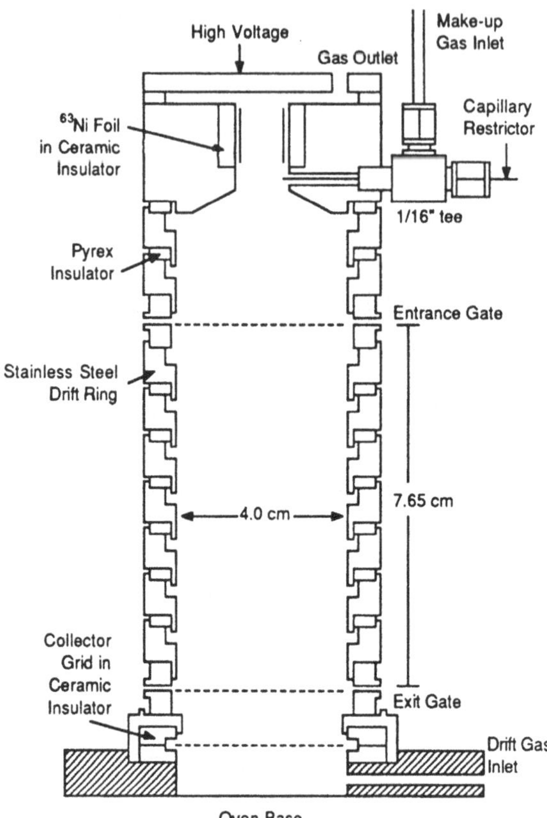

Figure 2.9 Schematic diagram of the ion-mobility detector. Reprinted from Morrissey *et al.* (1990), with permission.

after the entrance gate is shut. Thus, bromides can be detected selectively in the presence of chlorides. Although the formation of CO_2 clusters with the product ions may cause some problems, the detector can be used in SFC both with plain CO_2 and with modifiers (Eatherton *et al.*, 1986; Morrissey *et al.*, 1990; Huang *et al.*, 1991). Picogram to nanogram amounts can be detected.

2.7.10 Light-scattering detection

The light-scattering detector (LSD) is based on a non-selective detection principle for non-volatile compounds. The column effluent is nebulized, either by the natural expansion of a fluid or aided by another nebulizing gas (Figure 2.10). The droplets and particles formed in the nebulizer and in the drift tube are illuminated by a light beam from a laser or from another light source and the amount of scattered light is measured (Stolyhwo *et al.*,

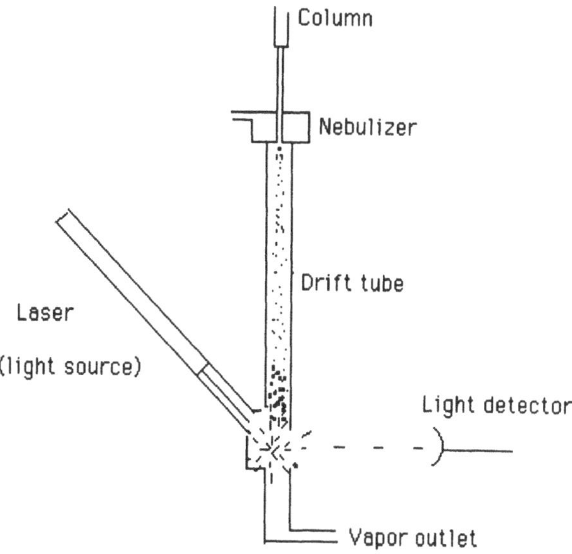

Figure 2.10 Schematic diagram of the light-scattering detector with a laser light source.

1983). The detector, which was originally made for HPLC, can be used in SFC without added nebulizing gas with standard size packed columns (Lafosse *et al.*, 1987) or with packed capillary columns (Hoffmann and Greibrokk, 1989), or with an inert gas for larger columns (Nizery *et al.*, 1989). The main advantage of the LSD is the freedom of choice of mobile phase, since this is removed in the nebulization process, together with volatile solutes. With modifiers added to carbon dioxide, polar compounds like ethoxylated alcohols and poly(ethylene glycols) have been separated and determined (Hagen *et al.*, 1991; Brossard *et al.*, 1992). Low-ng amounts can be determined. The response curve is nonlinear due to the complicated mechanisms of light scattering.

2.7.11 Mass spectrometric detection

As the most selective detector of all, the mass spectrometer plays an important role in SFC as well as in GC. Interfacing SFC to mass spectrometry (MS) is not as convenient as in GC–MS, but, compared to LC–MS, connections are relatively simple and better sensitivity is usually obtained (Smith *et al.*, 1982, 1984; Wright *et al.*, 1986). Density gradients can result in a loss of sensitivity, particularly to the heavier components, since the signal-to-noise ratio is reduced with increasing ion-source pressure. Ionization mechanisms include electron impact–charge transfer and chemical ionization, depending on the conditions. Most instruments are equipped with a quadrupole mass analyser, operating at lower potential and higher pressure than sector instruments, and at a lower price, for the bench-top instruments in particular (Lee and Henion, 1986). To enable a large portion of a packed column eluent to be handled by the mass spectrometer, a thermospray interface has been modified (Berry *et al.*, 1986; Raynor *et al.*, 1989). Sector instruments offer the advantage of higher resolution, with interfaces of slightly different designs (Kalinoski *et al.*, 1987; Huang *et al.*, 1988; Reinhold *et al.*, 1988; Kallio *et al.*, 1989; Huang and Morgan, 1990). Chemical ionization has been reported to result in sensitivities at least one order of magnitude better than electron impact (Lee *et al.*, 1988). Typical values for the minimum detectable quantities are in the ng area for a full scan and in the pg area for selected ion monitoring, but sub-picogram levels have been determined, e.g. for derivatized gangliosides (Merritt *et al.*, 1991). A more detailed overview of interfaces and ionization techniques in SFC–MS has been presented by Lee and Markides (1990).

References

Ashraf-Khorassani, M., Taylor, L.T. and Henry, R.A., (1989), Packed column SFC comparison of conventional and polymer-coated silica bonded phases. *Chromatographia*, **28**, 569–573.

Baastoe, M.B. and Lundanes, E., (1991), Capillary SFC of primary aliphatic amines using carbon dioxide and nitrous oxide as mobile phases. *J. Chromatogr.*, **558**, 458–463.

Baim, M.A. and Hill, H.H. Jr., (1982), Tunable selective detection for capillary GC by ion mobility monitoring. *Anal. Chem.*, **54**, 38–43.

Berg, B.E. and Greibrokk, T., (1989), The solvent venting injection technique in capillary SFC. *J High Resolut. Chromatogr.*, **12**, 322–326.

Berg, B.E. Flaaten, A.M., Paus, J. and Greibrokk, T., (1992), Extended use of solvent venting injection techniques for large sample volumes and coupled capillary columns in SFC. *J. Microcol. Sep.*, **4**, 227–232.

Berger, T.A. and Deye, J.F., (1990), Composition and density effects using methanol/carbon dioxide in packed column SFC. *Anal. Chem.*, **62**, 1181–1185.

Berry, A.J., Games, D.E. and Perkins, J.R., (1986), SFC and SFC–MS studies of some polar compounds. *J. Chromatogr.*, **363**, 147–158.

Blilie, A.L. and Greibrokk, T. (1985*a*), Gradient programming and combined gradient–pressure programming in SFC. *J. Chromatogr.*, **349**, 317–322.

Blilie, A.L. and Greibrokk, T., (1985*b*), Modifier effects on retention and peak shape in SFC. *Anal. Chem.*, **57**, 2239–2242.

Brossard, S., Lafosse, M. and Dreux, M., (1992), Comparison of ethoxylated alcohols and polyethylene glycols by high performance liquid chromatography or supercritical fluid chromatography using evaporative light-scattering detection. *J. Chromatogr.*, **591**, 149–157.

Chang, H.-C.K. and Taylor, L.T., (1990), The performance of electron-capture detection after capillary SFC. *J. Chromatogr. Sci.*, **28**, 29–33.

Chervet, J.P., Ursem, M., Salzmann, J.P. and Vannoort, R.W., (1989), Ultra-sensitive UV detection in micro separation. *J. High Resolut. Chromatogr.*, **12**, 278–281.

Chester, T.L., (1984), Capillary SFC with FID: reduction of detection artifacts and extension of detectable molecular weight range. *J. Chromatogr.*, **299**, 424–431.

Chester, T.L. and Innis, D.P., (1989), Sample loss and its control with internal-loop injection valves in SFC. *J. Microcol. Sep.*, **1**, 230–233.

Chester, T.L., Innis, D.P. and Owens, G.D., (1985), Separation of sucrose esters by capillary SFC/FID with robot-pulled capillary restrictors. *Anal. Chem.*, **57**, 2243–2247.

Cortes, H.J., Pfeiffer, C.D., Richter, B.E. and Stevens, T.S., (1987), Porous ceramic bed supports for fused silica packed capillary columns used in LC. *J. High Resolut. Chromatogr.*, **10**, 446–448.

Crow, J.A. and Foley, J.P., (1991), Formic acid modified carbon dioxide as a mobile phase in capillary SFC. *J. Microbiol. Sep.*, **3**, 47–57.

David, P.A. and Novotny, M., (1989), Analysis of steroids by capillary SFC with phosphorus-selective detection. *J. Chromatogr.*, **461**, 111–120.

Davies, I.L., Xu, B., Markides, K.E., Bartle, K.D. and Lee, M.L., (1989), Multidimensional open-tubular column SFC using a flow-switching interface. *J. Microcol. Sep.*, **1**, 71–84.

Dean, T.A. and Poole, C.F., (1989), Solventless injection for packed columns in SFC. *J. High Resolut. Chromatogr.*, **12**, 773–778.

de Weerdt, M., Dewaele, C., Verzele, M. and Sandra, P., (1990), Packing material activity in packed capillary column SFC. *J. High Resolut. Chromatogr.*, **13**, 40–46.

Eatherton, R.L., Morrissey, M.A., Siems, W.F. and Hill, H.H., Jr., (1986), Ion mobility detection after SFC. *J. High Resolut. Chromatogr.*, **9**, 154–160.

Farbrot Buskhe, A., Berg, B.E., Gyllenhaal, O. and Greibrokk, T., (1988), Splitless injection in capillary SFC. *J. High Resolut. Chromatogr.*, **11**, 16–20.

Farwell, S.O., Gage, D.R. and Kagel, R.A., (1981), Current status of prominent selective GC detectors: A critical assessment. *J. Chromatogr. Sci.*, **19**, 358–376.

Fields, S.M. and Grolimund, K., (1989), Evaluation of supercritical sulfur hexafluoride as a mobile phase for polar and non-polar compounds. *J. Chromatogr.*, **472**, 197–208.

Fields, S.M. and Lee, M.L., (1985), Effects of density and temperature on efficiency in capillary SFC. *J. Chromatogr.*, **349**, 305–316.

Fields, S.M., Markides, K.E. and Lee, M.L., (1988), UV detector for capillary SFC with compressible mobile phases. *Anal. Chem.*, **60**, 802–806.

Fjeldsted, J.C. and Lee, M.L., (1984), Capillary SFC. *Anal. Chem.*, **56**, 619A–628A.

Fjeldsted, J.C., Kong, R.C. and Lee, M.L., (1983*a*), Capillary SFC with conventional flame detectors. *J. Chromatogr.*, **279**, 449–455.

Fjeldsted, J.C., Richter, B.E., Jackson, W.P. and Lee, M.L., (1983*b*), Scanning fluorescence detection in capillary SFC. *J. Chromatogr.*, **279**, 423–430.

Foreman, W.T., Sievers, R.E. and Wenclawiak, B.W., (1988), SFC with redox chemiluminescence detection. *Fresenius Z. Anal. Chem.*, **330**, 231–234.

France, J.E. and Voorhees, K.J., (1988), Capillary SFC with UV multichannel detection of some pesticides and herbicides. *J. High Resolut. Chromatogr.*, **11**, 692–696.

French, S.B. and Novotny, M., (1986), Xenon, a unique mobile phase for SFC. *Anal. Chem.*, **58**, 164–166.

Fujimoto, C., Yoshida, H. and Jinno, K., (1987), Interfacing of ICP-AES to SFC for elemental detection. *J. Chromatogr.*, **411**, 213–220.

Fujimoto, C., Yoshida, H. and Jinno, K., (1989), The use of polar modifiers in microbore SFC combined with ICP spectrometry. *J. Microcol. Sep.*, **1**, 19–22.

Fuoco, R., Pentoney, S.L. and Griffiths, P.R., (1989), Comparison of sampling techniques for combined SFC and FT–IR with mobile phase detection. *Anal. Chem.*, **61**, 2212–2218.

Gemmel, B., Lorenschat, B. and Schmitz, F.P., (1989), Separation of oligomers of medium polarity by packed column SFC. *Chromatographia*, **27**, 605–610.

Gere, D.R., Board, R. and McManigill, D., (1982), SFC with small particle diamater packed columns. *Anal. Chem.*, **54**, 736–740.

Giddings, J.C., Myers, M.N. and King, J.W., (1969), Dense gas chromatography at pressures to 2000 atmospheres. *J. Chromatogr. Sci.*, **7**, 276–283.

Giorgetti, A., Pericles, N., Widmer, H.M., Anton, K. and Dätwyler, P., (1989), Mixed mobile phases and pressure programming in packed and capillary column SFC: A unified approach. *J. Chromatogr. Sci.*, **27**, 318–324.

Greibrokk, T., Blilie, A.L., Johansen, E.J. and Lundanes, E., (1984), New system for delivery of the mobile phase in SFC. *Anal. Chem.*, **56**, 2681–2684.

Greibrokk, T., Doehl, J., Farbrot, A. and Iversen, B., (1986), Mobile phase delivery in SFC. *J. Chromatogr.*, **371**, 145–152.

Greibrokk, T., Berg, B.E., Blilie, A.L., Doehl, J., Farbrot, A. and Lundanes, E., (1987), Techniques and applications in SFC. *J. Chromatogr.*, **394**, 429–441.

Greibrokk, T., Berg, B.E. and Johansen, H., (1988), Sample injection in capillary SFC. In

International Symposium on Supercritical Fluids, M. Perrut (Ed)., Nice, October 1988, pp. 425–430.
Guthrie, E.J. and Schwartz, H.E., (1986), Integral pressure restrictor for capillary SFC. *J. Chromatogr. Sci.*, **24**, 236–241.
Hagen, H.M., Eitrem Landmark, K. and Greibrokk, T., (1991), Separation of oligomers of polyethylene glycols by SFC. *J. Microcol. Sep.*, **3**, 27–31.
Harvey, M.C. and Stearns, S.D., (1983), HPLC sample injection using electric valve actuators. *J. Chromatogr. Sci.*, **21**, 473–477.
Hawthorne, S.B. and Miller, D.J., (1989), An almost on-column injector for capillary SFC. *J. Chromatogr. Sci.*, **27**, 197–202.
Hirata, Y., (1990), Column technology for packed capillary columns. *J. Microcol. Sep.*, **2**, 214–221.
Hirata, Y. and Nakata, F., (1984), SFC with fused-silica packed columns. *J. Chromatogr.*, **295**, 315–322.
Hirata, Y., Tanaka, M. and Inomata, K., (1989), Purged splitless injection in microcolumn SFC. *J. Chromatogr. Sci.*, **27**, 395–398.
Hirata, Y., Koshiba, H. and Maeda, T., (1990), Introduction of large sample volumes in capillary SFC. *J. High Resolut. Chromatogr.*, **13**, 619–623.
Hirata, Y., Kadota, Y. and Kondo, T., (1991), Technique for injecting very large volumes in capillary SFC. *J. Microcol. Sep.*, **3**, 17–25.
Hoffmann, S. and Greibrokk, T., (1989), Packed capillary SFC with mixed mobile phases and light-scattering detection. *J. Microcol. Sep.*, **1**, 35–40.
Huang, E.C., Jackson, B.J., Markides, K.E. and Lee, M.L., (1988), Direct heated interface probe for capillary SFC/double focusing MS. *Anal. Chem.*, **60**, 2715–2719.
Huang, H.-P. and Morgan, E.D., (1990), Analysis of azadirachtin by SFC. *J. Chromatogr.*, **519**, 137–143.
Huang, M.X., Markides, K.E. and Lee, M.L., (1991), Evaluation of an ion mobility detector for SFC with solvent-modified carbon dioxide mobile phases. *Chromatographia*, **31**, 163–167.
Hughes, M.E. and Fasching, J.L., (1985), Development and applications of a combined SFC with FTIR. *J. Chromatogr. Sci.*, **24**, 535–540.
Janssen, H.-G., Rijks, J.A. and Cramers, C.A., (1990), Flow rate control in pressure-programmed capillary SFC. *J. Microcol. Sep.*, **2**, 26–32.
Jinno, K., Hoshino, T., Hondo, T., Saito, M. and Senda, M., (1986), Identification of polycyclic aromatic hydrocarbons in extracts of diesel particulate matter by SFC coupled with an UV multichannel detector. *Anal. Chem.*, **58**, 2696–2699.
Jinno, K., Mae, H. and Fujimoto, C., (1990), Packed microcolumn SFC coupled with photodiode array UV detector and ICP. *J. High Resolut. Chromatogr.*, **13**, 13–17.
Johnson, C.C., Jordan, J.W., Taylor, L.T. and Vidrine, D.W., (1985), On-line SFC with FTIR spectrometric detection employing packed columns and a high pressure lightpipe flow cell. *Chromatographia,* **20**, 717–723.
Kalinoski, H.T., Udseth, H.R., Wright, B.W. and Smith, R.D., (1987), Analytical applications of capillary SFC-MS. *J. Chromatogr.*, **400**, 307–316.
Kallio, H., Laakso, P., Huopalahti, R. and Linko, R.R., (1989), Analysis of butter fat triacylglycerols by SFC/electron impact MS. *Anal. Chem.*, **61**, 698–700.
Kennedy, S. and Wall, R.J., (1988), Electron-capture detection of agrochemicals by SFC. **6**, 930–932.
Köhler, J., Rose, A. and Schomburg, G., (1988), Instrumentation for SFC systems: Different sampling and restriction designs. *J. High Resolut. Chromatogr.*, **11**, 191–197.
Koski, I.J., Markides, K.E., Richter, B.E. and Lee, M.L. (1991), Large volume sample introduction for open tubular SFC. In *13th Int. Symp. on Cap. Chrom.*, P. Sandra (Ed.), Riva del Garda, pp. 1628–1640.
Kuei, J.C., Markides, K.E. and Lee, M.L., (1987), Supercritical ammonia as mobile phase in capillary chromatography. *J. High Resolut. Chromatogr.*, **10**, 257–262.
Lafosse, M., Dreux, M. and Morin-Allory, L., (1987), Champs d'application d'un nouveau détecteur évaporatif à diffusion de lumière pour la chromatographie liquide hautes performances et la chromatographe en phase supercritique. *J. Chromatogr.*, **404**, 95–105.
Leach, R.A. and Harris, J.M., (1984), Supercritical fluids as spectroscopic solvents for thermooptical absorption measurements. *Anal. Chem.*, **56**, 1481–1487.

Lee, E.D. and Henion, J.D., (1986), Open tubular column SFC/MS on a benchtop mass spectrometer. *J. High Resolut. Chromatogr.*, **9**, 172–174.

Lee, E.D., Hsu, S.-H. and Henion, J.D., (1988), Electron-ionization-like mass spectra by capillary SFC/charge exchange MS. *Anal. Chem.*, **60**, 1990–1994.

Lee, M.L. and Markides, K.E., (Eds), (1990), *Analytical Supercritical Fluid Chromatography and Extraction*, Chromatography Conferences, Inc., Provo, UT, USA, pp. 264–285.

Lee, M.L., Xu, B., Huang, E.C., Djordjevic, N.M., Chang, H.-C.K. and Markides, K.E., (1989), Liquid sample introduction methods in capillary SFC. *J. Microcol. Sep.*, **1**, 7–13.

Levy, J.M. and Ritchey, W.M., (1986), Investigations of the uses of modifiers in SFC. *J. Chromatogr. Sci.*, **24**, 242–248.

Liu, Z., Farnsworth, P.B. and Lee, M.L., (1991), Sample introduction in capillary SFC using sequential density gradient focussing and solvent venting. *J. Microcol. Sep.*, **3**, 435–442.

Luffer, D.R. and Novotny, M.V., (1991), Capillary SFC and microwave-induced plasma detection of cyclic boronate esters of hydroxyl compounds. *J. Microcol. Sep.*, **3**, 39–46.

Luffer, D.R., Galante, L.J., David, P.A., Novotny, M. and Hieftje, G.M., (1988), Evaluation of a SFC coupled to a surface-wave-sustained microwave-induced plasma detector. *Anal. Chem.*, **60**, 1365–1369.

Markides, K.E., Lee, E.D., Bolick, R. and Lee, M.L., (1986), Capillary SFC with dual-flame photometric detection. *Anal. Chem.*, **54**, 740–743.

Martin, M. and Guiochon, G., (1978), Theoretical study of the gradient elution profiles obtained with syringe-type pumps in LC. *J. Chromatogr.*, **151**, 267–289.

Mathiasson, L., Jonsson, J.A. and Karlsson, L. (1989), Determination of nitrogen compounds by SFC using nitrous oxide as the mobile phase and nitrogen-sensitive detection. *J. Chromatogr.*, **467**, 61–74.

Merritt, M.V., Sheeley, D.M. and Reibhold, V.N., (1991), Characterization of glycosphingolipids by SFC–MS. *Anal. Biochem.*, **193**, 24–28.

Mol, J.G.J., Zegers, B.N., Lingeman, H. and Brinkman, U.A.Th., (1991), Packed-capillary SFC of pesticides using phosphorus-selective detection. *Chromatographia*, **32**, 203–210.

Morin, P., Caude, M., Richard, H. and Rosset, R., (1986), Carbon dioxide SFC–FTIR spectrometry. *Chromatographia*, **21**, 523–530.

Morrissey, M.A., Siems, W.F. and Hill, H.H., Jr., (1990), Ion mobility detection of polydimethylsilicone oligomers following SFC. *J. Chromatogr.*, **505**, 215–225.

Motley, C.B., Ashraf-Khorassani, M. and Long, G.L., (1989), Microwave-induced plasma as an elemental detector for packed column SFC. *Appl. Spectrosc.*, **43**, 737–741.

Moulder, R., Briggs, J., Cole, E.R., Bartle, K.D. and Clifford, A.A., (1991), SFC with electron capture detection. In *Proc. 13th Int. Symp. Cap. Chromatogr.*, P. Sandra (Ed.), Huethig, Heidelberg, pp. 1609–1613.

Mourier, P.A., Caude, M.H. and Rosset, R.H., (1985*a*), Rétention, sélectivité et efficacité en chromatographie en phase supercritique. *Analusis*, **13**, 299–311.

Mourier, P.A., Eliot, E., Caude, M.H., Rosset, R.H. and Tambute, A.G., (1985*b*), Supercritical and subcritical fluid chromatogrpahy on a chiral stationary phase for the resolution of phosphine oxide enantiomers. *Anal. Chem.*, **57**, 2819–2823.

Nizery, D., Thiebaut, D., Claude, M., Rosset, R., Lafosse, M. and Dreux, M., (1989), Improved evaporative light-scattering detection for SFC with carbon dioxide–methanol mobile phases. *J. Chromatogr.*, **467**, 49–60.

Novotny, M., Springston, S.R., Peaden, P.A., Fjeldsted, J.C. and Lee, M.L. (1981), Capillary SFC. *Anal. Chem.*, **53**, 407A–417A.

Novotny, M., Lee, M.L., Peaden, P.A., Fjeldsted, J.C. and Springston, S.R., (1985), Open-tubular SFC. U.S. Pat. no. 4 479 380.

Olesik, S.V., French, S.B. and Novotny, M., (1984), Development of capillary SFC/FTIR. *Chromatographia*, **18**, 489–495.

Olesik, J.W. and Olesik, S.V., (1987), Supercritical fluid-based sample introduction for inductively coupled plasma atomic spectrometry. *Anal. Chem.*, **59**, 796–799.

Payne, K.M., Tarbet, B.J., Bradshaw, J.S., Markides, K.E. and Lee, M.L., (1990), Simultaneous deactivation and coating of porous silica particles for microcolumn SFC. *Anal. Chem.*, **62**, 1379–1384.

Peaden, P.A. and Lee, M.L., (1983), Theoretical treatment of resolving power in open tubular column SFC. *J. Chromatogr.*, **259**, 1–16.

Pekay, L.A. and Olesik, S.V., (1989), SFC/FPD: Determination of high MW compounds. *Anal. Chem.*, **61**, 2616–2624.

Pekay, L.A. and Olesik, S.V., (1990), Evaluation of sulfur chemiluminescence detection in SFC. *J. Microcol. Sep.*, **2**, 270–277.

Porter, N.L., Richter, B.E., Bornhop, D.J., Later, D.W. and Beyerlein, F.H., (1987), Effects of fluid filling techniques on reproducibility in capillary SFC. *J. High Resolut. Chromatogr.*, **10**, 477–478.

Rawdon, M.G., (1984), Modified FID for SFC. *Anal. Chem.*, **56**, 831–832.

Raynor, M.W., Bartle, K.D., Davies, I.L., Williams, A., Clifford, A.A., Chalmers, J.M. and Cook, B.W., (1988), Polymer additive characterization by capillary SFC/FTIR microspectrometry. *Anal. Chem.*, **60**, 427–433.

Raynor, M.W., Kithinji, J.P., Bartle, K.D., Games, D.E., Mylchreest, I.C., Morgan, E.D. and Wilson, I.D., (1989), Packed column SFC for the analysis of phytoecdysteroids from *Silene nutans* and *Silene otites*. *J. Chromatogr.*, **467**, 292–298.

Raynor, M.W., Bartle, K.D., Clifford, A.A. and Cook, B.W., (1990), Stopped-flow FTIR detection in capillary SFC. *J. Microcol. Sep.*, **2**, 300–303.

Reinhold, V.N., Sheeley, D.M., Kuei, J. and Her, G.-R., (1988), Analysis of high molecular weight samples on a double-focusing magnetic sector instrument by SFC/MS. *Anal. Chem.*, **60**, 2719–2722.

Richter, B.E., Knowles, D.E., Andersen, M.R., Porter, N.L., Campbell, E.R. and Later, D.W., (1988), Reproducibility in capillary SFC: Comparison of injection techniques. *J. High Resolut. Chromatogr.*, **11**, 29–32.

Richter, B.E., Bornhop, D.J., Swanson, J.T., Wangsgaard, J.G. and Andersen, M.R., (1989), GC detectors in SFC. *J. Chromatogr. Sci.*, **27**, 303–308.

Rivière, B., Mermet, J.-M. and Deruaz, D., (1988), Spectroscopic evaluation of a carbon dioxide and a helium–carbon dioxide microwave-induced plasma (Surfatron). *J. Anal. Atomic Spectrom.*, **3**, 551–555.

Saito, M., Yamauchi, Y. and Kashiwazaki, H., (1988), New pressure regulating system for constant mass flow SFC and physico-chemical analysis of mass-flow reduction in pressure programming by analogous circuit model. *Chromatographia*, **25**, 801–805.

Saito, T. and Takeuchi, M., (1989), Development of semi-micro supercritical fluid chromatograph and its application. *Jeol News*, **25A**, 14–17.

Sanagi, M.M. and Smith, R., (1988), The emergence and instrumentation of SFC. In *Supercritical Fluid Chromatography* (R. Smith, Ed.), Royal Society of Chemistry, London, pp. 43–52.

Saunders, C.W. and Taylor, L.T., (1989), The effect of elevated temperature on conventional stationary phases used in SFC. *Chromatographia*, **28**, 253–257.

Schaefer, B.A., (1972), Response of the flame ionization detector to oxygen and nitrous oxide. *J. Chromatogr. Sci.*, **10**, 111–120.

Schoenmakers, P.J. and Uunk, L.G.M., (1987), Effects of the column pressure drop in packed-column SFC. *Chromatographia*, **24**, 51–57.

Schomburg, G. and Roeder, W., (1989), Sampling techniques for SFC using peak focusing by pressure and/or temperature programming. *J. High Resolut. Chromatogr.*, **12**, 218–225.

Schomburg, G., Behlau, H., Haüsig, U., Hoening, B. and Roeder, W., (1989), Quantitation in miniaturized GC and SFC systems. *J. High Resolut. Chromatogr.*, **12**, 142–148.

Schwartz, H.E. and Brownlee, R.G., (1986), Hydrocarbon group analysis of gasolines with microbore SFC and FID. *J. Chromatogr.*, **353**, 77–93.

Scott, R.P.W. and Simpson, C.F., (1982), Determination of the extra column dispersion occurring in the different components of a chromatographic system. *J. Chromatogr. Sci.*, **20**, 62–66.

Shafer, K.H. and Griffiths, P.R., (1983), On-line SFC/FTIR. *Anal. Chem.*, **55**, 1939–1942.

Shafer, K.H., Pentoney, S.L. and Griffiths, P.R., (1984), SFC/diffuse reflectance FTIR. *J. High Resolut. Chromatogr.*, **7**, 707–709.

Shah, S. and Taylor, L.T., (1989), On-line SFC–FTIR analysis of agriculturally-related compounds. *J. High Resolut. Chromatogr.*, **12**, 599–603.

Shen, W.-L., Vela, N.P., Sheppard, B.S. and Caruso, J.A. (1991), Evaluation of ICP–MS as an elemental detector for SFC. *Anal. Chem.*, **63**, 1491–1496.

Sie, S.T., van Beersum, W. and Rijinders, G.W.A. (1966), High pressure GC and chromatography with supercritical fluids. I. The effect of pressure on partition coefficients in GLC with carbon dioxide as a carrier gas. *Sep. Sci.*, **1**, 459–490.

Simons, J.K., Sin, C.H., Zabriskie, N.A., Fields, S.M., Lee, M.L. and Goates, S.R. (1989), SFC–supersonic jet spectroscopy: I. Microcolumns and direct expansions. *J. Microcol. Sep.*, **1**, 200–206.

Simpson, R.C., Gant, J.R. and Brown, P.R., (1986), Modification of conventional HPLC equipment for use in carbon dioxide SFC. *J. Chromatogr.*, **371**, 109–119.

Skelton, R.J., Jr., Markides, K.E., Farnsworth, P.B. and Lee, M.L., (1988), Multi-element selective radio frequency plasma detector for capillary gas chromatography. *J. High Resolut. Chromatogr.*, **11**, 75–82.

Smith, R.D., Felix, W.D., Fjeldsted, J.C. and Lee, M.L., (1982), Direct fluid injection interface for capillary SFC–MS. *J. Chromatogr.*, **247**, 231–243.

Smith, R.D., Udseth, H.R. and Kalinoski, H.T., (1984), Capillary SFC/MS with electron impact ionization. *Anal. Chem.*, **56**, 2971–2973.

Stolyhwo, A., Colin, H. and Guiochon, G., (1983), Use of light scattering as a detector principle in LC. *J. Chromatogr.*, **265**, 1–18.

Tuominen, J.P., Markides, K.E. and Lee, M.L., (1991), Optimization of internal valve injection in open tubular SFC. *J. Microcol. Sep.*, **3**, 229–239.

West, W.R. and Lee, M.L., (1986), Evaluation of thermoionic detector for capillary SFC of nitrated polycyclic aromatic compounds. *J. High Resolut. Chromatogr.*, **9**, 161–167.

Wieboldt, R.C., Adams, G.E. and Later, D.W., (1988), Sensitivity improvement in infrared detection for SFC. *Anal. Chem.*, **60**, 2422–2427.

Wright, B.W., Kalinoski, H.T., Udseth, H.R. and Smith, R.D., (1986), Capillary SFC/MS. *J. High Resolut. Chromatogr.*, **9**, 145–153.

Yang, F.J., (1989), In *1989 Symposium/Workshop on SFC*, K.E. Markides and M.L. Lee (Eds), Comps., Brigham Young University Press, Provo, UT, USA.

Yonker, C.R. and Smith, R.D., (1986), Study of retention processes in capillary SFC with binary fluid mobile phases. *J. Chromatogr.*, **361**, 25–32.

Zhang, L., Carnahan, J.W., Winans, R.E. and Neill, P.H., (1991), SFC with a helium microwave-induced plasma for chlorine-selective detection. *Anal. Chem.*, **63**, 212–216.

3 Instrumentation for supercritical fluid extraction

J.R. DEAN and M. KANE

3.1 Introduction

Sample preparation frequently involves the conversion of a solid sample into a liquid prior to analysis (Majors, 1991*a*). This can be achieved by using liquid–solid extraction techniques for isolating the analyte from its matrix. Traditionally, Soxhlet extraction can be used effectively for this purpose. Soxhlet extraction involves the repeated distillation of the solid sample in a suitable solvent. The technique is far from ideal, requiring the use of large volumes of organic solvents, and can be rather slow for analyte recovery. Additionally, after extraction there is the requirement to dispose of the organic solvent in an appropriate manner. However, despite these inadequacies Soxhlet extraction remains a useful and reliable sample-preparation technique in the analytical laboratory. Recent advances have seen the introduction of Soxhlet extraction in the form of Soxtec, a technique that uses less of the organic solvent, and has the facility for multiple samples and faster extraction times. The complete removal of organic solvents and faster extraction times would be considered a major advance for sample preparation. The mid 1980s saw the arrival of analytical supercritical fluid extraction (SFE) (King, 1989; Hawthorne, 1990; Lee and Markides, 1990; Engelhardt and Gross, 1991; Majors, 1991*b*). SFE has (i) potential for reduced extraction times, (ii) controllable extraction conditions, (iii) potential for fractionation, (iv) reduced risk of contamination, (v) compatibility with on-line methods of analysis (e.g. chromatography), (vi) flexibility for off-line analysis (e.g. spectrophotometry), and (vii) possibility of class-selective extraction by appropriate choice of conditions. The properties and conditions for obtaining the supercritical state have been described in chapter 1. It is worthwhile reviewing the suitability of supercritical carbon dioxide (Table 3.1). From this it is easy to see why carbon dioxide is the most common of the supercritical fluids in operation in the analytical laboratory.

However, the major limitation of carbon dioxide is its inability to extract polar compounds at typical working pressures (80–600 atm). This situation can be remedied by increasing the polarity of the supercritical carbon dioxide by addition of small amounts of polar compounds, e.g. acetonitrile, methanol, etc. The practical and sometimes problematic methods of adding these modifiers will be considered later. Alternatively, the use of a

Table 3.1 Benefits of supercritical carbon dioxide

(a)	Moderate critical pressure (73.8 bar)
(b)	Low critical temperature (31.1 °C)
(c)	Low toxicity and reactivity
(d)	High purity at low cost
(e)	Useful for extractions at temperatures <150 °C
(f)	Ideal for extraction of thermally labile compounds
(g)	Ideal extractant for non-polar species, e.g. alkanes
(h)	Reasonably good for moderately polar species, e.g. PAHs, PCBs
(i)	Can directly vent to atmosphere
(j)	Little opportunity for chemical change in absence of light and oxygen
(k)	Gas at room temperature, allows for coupling to gas chromatography and SFC

supercritical fluid with a higher solvent strength is potentially viable (Li *et al.*, 1990). However, practical limitations frequently prevent this in reality. Supercritical ammonia has the potential for extracting more polar compounds but has a tendency to dissolve pump seals, thereby making it difficult to pump. It is also chemically reactive and too dangerous for routine use. The practical variables that will ultimately affect the class selective supercritical fluid extraction efficiency are outlined in Table 3.2.

Table 3.2 Supercritical fluid extraction operation variables

Supercritical fluid
Choice
Flow rate
Pressure (density)
Temperature
Modified supercritical fluid
Selection of appropriate modifier
Concentration of modifier
Sample
Sample condition, e.g. water content, pH
Size
Matrix
Particle size
Analyte type, e.g. polarity, molecular weight
Extraction cell
Geometry
Agitation
Dead volume
Cell size
Extraction time
Extraction cell temperature
Restrictor
Restrictor type, i.e. fixed or variable
Sample accumulation
Mode of sample accumulation, e.g. on-line or off-line; dry collection or in solvent

3.2 SFE instrumentation

The basic components of an SFE system are shown in Figure 3.1. The main features of the instrumentation are a pump, an extraction vessel, a restrictor and an analyte collection device. In addition, a high-purity gas supply (free from water, halocarbons and hydrocarbons) of carbon dioxide is required. High-purity SFE-grade carbon dioxide, dip-tube fitted, with or without an overpressure of helium (Gorner *et al.*, 1990) is now available from commercial outlets worldwide. Typical analytical data from one supplier of a range of carbon dioxide purities are provided in Table 3.3. Failure to use high-purity carbon dioxide may result in the pre-concentration of trace impurities which are collected and analysed. The gas is pressurized to above the critical pressure by using a pump. The pump used must be able to generate high pressure, typically 3750–10 000 psi, and supply a constant flow rate from $\mu l\ min^{-1}$ to $ml\ min^{-1}$ reproducibly. Typically, a syringe pump or a reciprocating pump is used. Both types are suitable for SFE. In order to operate with these pumps some modifications are required, owing to the different compressibilities of liquids and supercritical fluids.

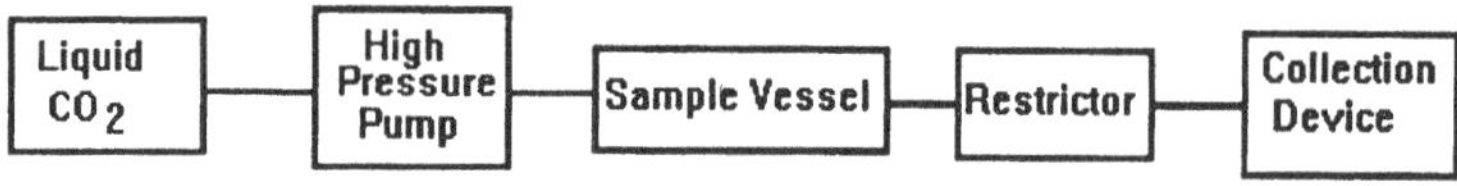

Figure 3.1 Block diagram of main components in SFE.

3.2.1 Reciprocating pumps

A reciprocating pump operates by driving a small motor-driven piston in and out of a hydraulic chamber. On the backward stroke, the inlet check valve opens and solvent is sucked into the chamber from the solvent reservoir. On the forward stroke, the outlet valve opens and carbon dioxide is pumped out. Dual-headed pumps consist of identical hydraulic chambers and pistons that are operated 180° out of phase. This has the benefit of delivering a smoother flow of carbon dioxide because in a dual-headed system one hydraulic chamber is filling whilst the other is pumping out the carbon dioxide. This provides the major advantage of a reciprocating pump, that of a continuous supply of liquid carbon dioxide. However, as the supercritical fluid has a much lower viscosity than a typical liquid, there is a tendency for the pump head to leak if it is not cooled. In SFE it is therefore essential to cool the pump head in order to pump the carbon dioxide. This can be achieved by using low-cost cryogenic-grade carbon dioxide or an ethylene glycol mixture pumped using a recirculating bath.

Table 3.3 Typical analytical data from a carbon dioxide supplier[a]

Typical impurities analysis	Industrial grade without dip tube	Industrial grade with dip tube	High purity grade with dip tube	SFC grade	SFE–SFC grade
Carbon dioxide	>99.5%	>99.99%	>99.997%	99.995%	>99.999%
Nitrogen	<500 ppm	<50 ppm	<10 ppm	10 ppm	
Oxygen	<200 ppm	<20 ppm	<10 ppm	7 ppm	
Water	<10 ppm	<10 ppm	<5 ppm	3 ppm	<200 ppb
Carbon monoxide	<1 ppm[b]	<1 ppm[b]	<1 ppm[b]	1 ppm	
Total hydrocarbons[c]	<10 ppm	<10 ppm	<5 ppm	3 ppm	<100 ppb
Total sulphur[d]	<0.1 ppm	<0.1 ppm	<0.1 ppm	0.1 ppm	
Total halocarbons[e]					<0.1 ppb

Notes: [a]Air Products plc, Weston Road, Crewe CW1 1DF, UK (reproduced with permission); [b]limits of detectability; [c]expressed as methane; [d]expressed as sulphur dioxide; [e]relative to Aldrin electron-capture detector response.

Even so, it is often necessary to add organic solvents as modifiers to the carbon dioxide to increase the polarity of the supercritical fluid mixture. This is not easily done by using a single reciprocating pump but requires the use of a second pump. The second pump, which does not require additional pump-head cooling, can be used to supply the organic modifier. The carbon dioxide and organic modifier can be premixed in a T-piece to varying proportions to affect extraction. This allows a greater amount of control and freedom for selecting the appropriate organic modifier for a particular extraction. A single pump can be used to provide the addition of an organic modifier, provided that the modifier is added either directly to the sample as a spike or by the use of a liquid carbon dioxide cylinder doped with a pre-specified amount of modifier. Obviously, the latter does not allow for flexibility in the choice of modifier or the concentration of modifier to be used. There is also the problem of obtaining a representative carbon dioxide:modifier ratio over the lifetime of the cylinder (Schoenmakers, 1991).

3.2.2 Syringe pumps

A syringe pump operates by positive displacement of the carbon dioxide from the chamber by using a variable-speed stepper motor to turn a screw which drives a piston. The positive displacement of carbon dioxide eliminates potential cavitation and repressurization inconsistencies. This provides a pulseless flow of carbon dioxide. Syringe pumps, in contrast to the reciprocating pump, have the major advantage of requiring no additional cooling of the liquid carbon dioxide. This is because the carbon dioxide is liquefied by pressure, not temperature. However, the major disadvantage of a syringe pump is that the carbon dioxide capacity is limited to the volume of the solvent chamber, typically 50–500 ml. Also, the limited volume ensures that the syringe must be filled on a regular basis and repressurized. Addition of modifier can be achieved by the use of a second syringe pump connected in parallel, or by adding a pre-selected amount of modifier to the pump prior to filling with carbon dioxide. This latter case may well provide a variable carbon dioxide:modifier ratio similar to the modified cylinders discussed above.

3.2.3 *Other pump modules*

In addition to the more common reciprocating and syringe pumps, other pumping systems have been used. This has been necessitated by the increase in cost brought about by the need to cool the pump head of a reciprocating pump. Pawliszyn (1990) described an inexpensive fluid delivery system for SFE based on high-pressure vessels without the need for additional cooling (Figure 3.2). The vessels are constructed of stainless steel with a volume of about 200 ml and are able to withstand pressures in excess of 1400 atm. The volume of the vessel allows for about 5 h of continuous supercritical flow at about 150 ml min^{-1} carbon dioxide with the 20 μm restrictor. For the safe operation of such a system each vessel is fitted with a valve with discs that rupture at 7000 atm. Narrow-bore copper tubing welded to the outside surface of the vessel provides cooling to the incoming fluid during servicing. The fluid is cooled to below its critical point with liquid nitrogen. Heating of the high-pressure vessel is maintained by a beaded heater. The pressure is monitored electronically by a pressure transducer and visually with a pressure gauge. The temperature and pressure in the vessel are controlled and monitored electronically. Two high-pressure vessels are operated in parallel allowing continuous supercritical fluid delivery by the use of a control valve. In operation, one vessel is supplying a continuous flow of supercritical fluid while the other vessel is being serviced and stabilized at the extraction pressure. Servicing of the high-pressure vessel involves a cooling process, refilling with carbon dioxide and heating to achieve the required pressure. Upon depletion of the in-

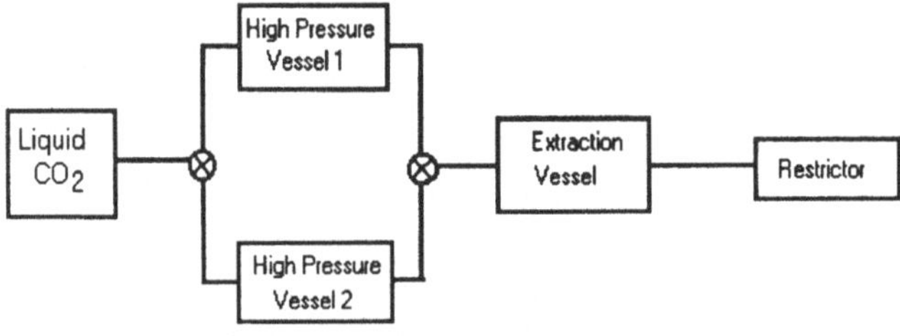

Figure 3.2 Continuous-fluid delivery using two high-pressure vessels. Reproduced with permission from Pawliszyn (1990).

line vessel the valve is switched to the second vessel allowing continuity in supercritical supply. The depletion in the contents of the vessel is noted by an increase in its temperature. The potential for miniaturization of the fluid-delivery system exists by replacing the high-pressure vessels with stainless-steel tubing. This would lead to difficulties associated with limited sample size and the need for continual refilling of the two pumps. It would, however, provide advantages associated with the safety aspects of using high-pressure vessels, and the ability to extract small amounts of sample and deposit them directly on to a capillary column.

The choice of modifier and method of addition are important, as it is used to increase the polarity of the supercritical carbon dioxide and thus to increase the extraction efficiency. The methods normally used to add modifiers, as a doped cylinder or by mixing with the aid of a second pump, were described above. Hawthorne *et al.* (1991) have described a method that allows for the preparation of modified carbon dioxide or nitrous oxide without exposing the pump to the organic modifier, a situation not previously possible using any of the previous methods of modifier addition. The benefits of this system (Figure 3.3) are that contamination of the pump by the modifier is eliminated and that several different modifiers can rapidly be evaluated. The system consists of a syringe pump and a four-port valve for diverting the flow of the carbon dioxide or nitrous oxide into the extraction cell either directly or through the modifier vessel. The modifier vessel is constructed from stainless steel with a volume of 9.5 ml. The temperature of the modifier vessel is maintained at 60 °C by placing it in a gas chromatography (GC) oven. Extractions are performed by press-

Figure 3.3 Device for the preparation of modified supercritical carbon dioxide. Reproduced with permission from Hawthorne *et al.* (1991).

urizing the sample cell with the pure carbon dioxide or nitrous oxide for about 3 min, then rotating the valve to allow flow of the carbon dioxide or nitrous oxide into the modifier vessel. The saturated carbon dioxide or nitrous oxide leaving the modifier vessel is then directed by the valve into the extraction cell. Carbon dioxide mixtures saturated with a range of modifiers (propylene carbonate, 2-methoxyethanol, acetic acid, 1-butanol and methanol) were evaluated at 60 °C using the modifier vessel, 50 mg samples of linear alkyl benzenesulphonates and a flow of 1.2 ml min^{-1} carbon dioxide for 15 min. The approximate concentration of the modifiers in the carbon dioxide saturated at 60 °C ranged from 15% for propylene carbonate to 40% for methanol. The system was evaluated at two extraction temperatures, 65 and 125 °C. Whereas no detectable recovery of the linear alkyl benzenesulphonates was observed using carbon dioxide or nitrous oxide individually, a near 100% recovery was observed using the methanol-modified carbon dioxide at 125 °C. The uncertainty as to whether some of the modifiers used were at supercritical conditions at the lower extraction temperature prevents immediate discrediting of the suitability of the modifier. However, the existence of suitable phase diagrams (Brunner *et al.*, 1987) for methanol–carbon dioxide mixtures shows that extractions with methanol were supercritical at each temperature tested.

3.2.4 Sample cell

The pressurized carbon dioxide is used to extract the analyte from its matrix contained in a sample cell. Sample cells typically range in size from 150 μl to 50 ml and are constructed from stainless steel or similar inert material. It is essential, however, that the sample vessels are rated for the maximum working pressure of the system. Temperature control of the sample cell is usually maintained by placing the vessel in a chromatographic oven or heating block. A suitable time interval should be allowed to enable the sample cell and contents to reach its pre-set temperature, therefore allowing for thermal lag prior to extraction. The type of sample cell used should ideally allow for quick change without the need for wrenches in order to achieve a high pressure seal. This may be achieved by using vessels that require some minimal personal interjection, such as finger-tight connections, or direct insertion. Currently available sample cell designs fall into two categories. The predominant one is the flow-through design, Figure 3.4(a), in which supercritical carbon dioxide passes continuously over the sample. In contrast, the less popular design for analytical SFE (Figure 3.4(b)) allows the sample matrix to be surrounded by supercritical carbon dioxide and the headspace atmosphere to be sampled. The latter type has been used for larger volume samples, typically 40 ml.

The effect of extractor-cell geometry has been investigated in a series of papers by K.G. Furton, and co-workers (Furton and Rein, 1991*a*,*b*; Rein

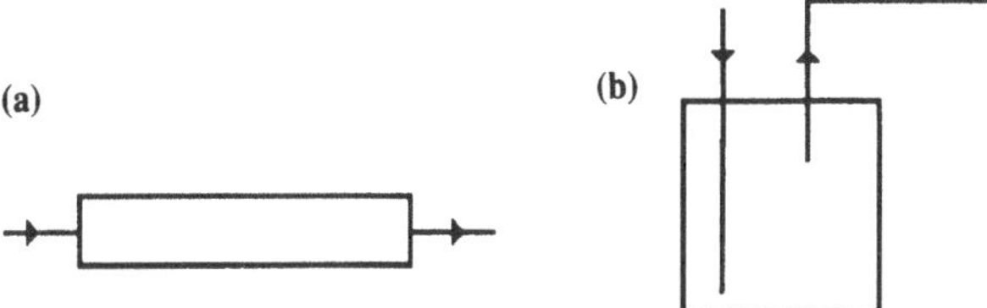

Figure 3.4 Types of extraction vessel used in analytical SFE: (a) flow-through cell; (b) headspace-sampling cell.

et al., 1991) on the extraction of polycyclic aromatic hydrocarbons (PAHs) from octadecyl-bonded sorbents. High-purity carbon dioxide was pumped using a syringe pump through a micro-extractor cell in an oven and then depressurized using linear restrictors fabricated from fused-silica tubing. Samples (0.5 g) of the standard packing were weighed directly into the extraction cells. Two different geometries of extraction cells were used: (a) short geometry, with 1.0 × 1.0 cm dimensions (i.d. × length of bed); and (b) long geometry, 0.37 × 7.3 cm dimensions (i.d. × length of bed). The short-geometry cell had a 1:1 diameter-to-length ratio, whereas the long geometry cell had a 1:20 diameter-to-length ratio. Four PAHs with increasing fused-ring numbers from four to seven were compared (pyrene, perylene, benzo[*ghi*]perylene and coronene) using constant extraction conditions that yielded less than 100% recoveries of the analytes. The relative increase in percentage recoveries using 1:1 versus 1:20 extraction-cell geometries were 33, 109, 137 and 206% for pyrene, perylene, benzo[*ghi*]-perylene and coronene, respectively. These results corresponded to an almost linear relationship with the number of fused rings in the PAHs studied. In contrast to these results were those obtained when the density of the supercritical fluid was increased by 0.2 g ml^{-1}. The relative increases in percentage recoveries for the analytes described above were 353, 311, 270 and 236%, respectively. This was the reverse order to the results obtained with cell geometry. The results for the changing-density experiment are not so surprising considering that the supercritical density is considered to be a major variable. What is surprising is the similar magnitude of effects observed with changing extraction-cell geometry. Obviously, the interaction of all the operating parameters can have a role in obtaining efficient class-selective extraction. The use of chemometric methods of evaluation could no doubt contribute significantly to this particular field of enquiry (Kane *et al.*, 1993).

The packing of the sample in the extraction cell can be influential on the rate of extraction. A study by Richards and Campbell (1991) ensured that the soil sample for extraction was packed so as to allow the minimum of dead volume to remain in the extraction cell. This was achieved by using glass beads (80 μm) placed at either end of the sample-containing flow-

Figure 3.5 Schematic diagram of the Hewlett-Packard HP7680A SFE system, showing capability for static (bypass valve open, chamber valve closed) and dynamic (bypass valve closed, chamber valve open) fluid flow. Reproduced with permission of W. Pipkin, Hewlett-Packard Ltd, Avondale, PA, USA.

through extraction cell. Additionally, the glass beads prevented blockage of the extraction-vessel outlet frit.

Extractions can be done in three distinct ways: (i) static, (ii) dynamic, or (iii) recirculating mode. In the static mode (Figure 3.5; bypass valve open, chamber valve closed) the sample-containing cell is filled with supercritical carbon dioxide; this allows for equilibration of the sample with the supercritical fluid prior to depressurization and subsequent collection. In the dynamic mode (Figure 3.5; bypass valve closed, chamber valve open), fresh supercritical fluid is continuously passed over the sample for a prespecified time. In practice, for moderately polar analytes, a combination of static and dynamic extraction is used with supercritical carbon dioxide. Finally, in the recirculating mode (Figure 3.6), a supercritical fluid is repeatedly passed over the sample for a period of time. In development work it is advantageous to assess the current status of extraction by using either the static or the recirculating mode. This could be done successfully by using an in-line monitoring system. Typically, this may be a single- or variable-wavelength ultraviolet–visible spectrophotometer with high-pressure sampling cell. Obviously, other forms of spectroscopy including

Figure 3.6 Schematic diagram of the LDC sample preparation accessory for recirculating of supercritical fluid. Reproduced with permission from Subra and Boissinot (1991).

Fourier transform infrared (FT-IR) may be useful for this purpose, as well as specific detectors such as nitrogen phosphorus detectors (NPDs).

A variety of systems is available that allows for single or multiple extractions to be done sequentially or simultaneously. The advantages to be achieved by multisample simultaneous extractions are obvious, relating to enhanced productivity and lower costs per unit time. However, what may not be clear to the unwary are the vagaries associated with the understanding of SFE and its associated parameters (Table 3.2). A firm theoretical understanding that allows the prediction of the appropriate operating conditions for compounds with similar structural features is slow in appearing (Bartle *et al.*, 1990, 1991). This must be the prelude to acceptance of the technique as an essential prerequisite for all analytical laboratories. Currently, eight manufacturers are supplying commercial SFE instrumentation on the world market (Table 3.4). Details of currently available instrumentation are given in Table 3.5.

Table 3.4 Suppliers of commercial supercritical fluid extractors

Carlo Erba Strumentazione SPA, Milan, Italy
Computer Chemical Systems (CCS), Avondale, Pennsylvania, USA
Dionex Corporation, Sunnyvale, California, USA
Hewlett-Packard Ltd, Avondale, Pennsylvania, USA
Isco Inc., Lincoln, Nebraska, USA
Jasco International Co. Ltd, Tokyo, Japan
LDC Analytical, Riviera Beach, Florida, USA
Suprex Corporation, Pittsburgh, Pennsylvania, USA

Table 3.5 Commercial supercritical fluid extractors

	Hewlett-Packard	Suprex	Jasco	Isco	CCS	Dionex	Carlo Erba	LDC
Model(s) available	7680A	SFE50, SFE50M, MPS225 and PrepMaster	Super 200	SFX 2-10	3200, 3100-100, 3200S and 3200P	SFE-703	SFE 30	SPA
Free-standing or hyphenation	Off-line	Both available	Off-line	Off-line	Both available	Off-line	On-line SFC with possibility for off-line	Off-line plus direct coupling to HPLC or GC
Pump	Piston	Piston or syringe 250 ml capacity	Piston	Syringe, 260 ml capacity	Both, depending on model	Pneumatic	Syringe, 150 ml capacity	Piston
Maximum operating pressure of system (psi)	5560	7350 (standard), 10 000 (optional)		3750, 7500 and 10 000 (depending on model)	6000 and 10 000	10 000	5800	5000
Pump cooling	CO_2	None	Ethylene glycol	None	Yes	None	None	Ethylene glycol
Addition of modifier	Direct to sample or premixed cylinder	Add to pump or use second pump	Second pump	Add to pump or use second pump	Modifier mixer	Direct to sample or premixed cylinder	Add to pump or use second pump	Direct to sample or premixed cylinder

Extraction cell sizes	1.5 or 7 ml	0.15–50 ml (depends on model)	1, 10 and 50 ml	0.5, 2.5 and 10 ml	1.4, 4, 10, 20, 32 and 100 ml (depends on model)	0.5–32 ml	0.4 or 1.5 ml (larger cells available, 30–50 ml)	5 ml
Static, dynamic or recirculating extraction	Static and dynamic	Static or dynamic	Dynamic only	Dynamic	Static or dynamic	Dynamic	Dynamic	Dynamic or recirculating
Sample extraction	Sequential	Mainly sequential	Sequential	Simultaneous (2) but can add on 4 more	Mainly sequential but simultaneous (6) available	Simultaneous (8)	Sequential	Sequential
Depressurization	Variable restrictor nozzle	Restrictor	Back-pressure regulator	Restrictor	Restrictor	Restrictor	Restrictor	Orifice
Extract collection	Chromatographic column (ODS) or stainless steel balls	Cryogenic trap available	Vial with or without solvent	Solvent or empty vessel	Cryogenic trap or in solvent	Solvent or empty vessel	Solvent, empty vessel or solid support	Solvent or empty vessel
Other features	Computer control	On-line capability to GC, SFC and HPLC	In-line UV monitoring	Top-loading sample cells	On-line coupling to GC or SFC available	Heated restrictor	In-line monitoring using GC and HPLC detectors	In-line UV monitoring

3.3 Depressurization systems

3.3.1 Restriction devices

One of the most crucial parts of the extraction process is the depressurization step, where the supercritical solvent returns to atmospheric conditions, allowing the extracted analytes either to be collected (off-line SFE) or to be directly analysed (on-line SFE). The technology which facilitates this step is shared with supercritical fluid chromatography (SFC) and can be split into two general categories: (i) simple fixed restrictors; and (ii) mechanical or electronically controlled restrictors. Before the two types of restrictor are discussed it is important to understand clearly the need for restriction in the extraction process.

Supercritical conditions are maintained in the fluid line by temperature and pressure conditions which are above the critical point (P_c). The pressure in the system is created by the back pressure induced by the restrictor device. In the case of a CO_2-based extraction fluid the pump delivers the CO_2 in liquid form at a constant flow rate, and supercritical conditions are established in the heated extraction chamber. The restrictor not only determines the back pressure, but also controls the mass flow rate of supercritical fluid flowing through the system. The smaller the orifice of the restrictor then the greater the back pressure in the line and the greater the mass flow rate. With variable restrictors the size of orifice can change and therefore the flow at the pump is independent of the pressure of the system. With a fixed restrictor an increase in flow at the pump will mean an increase in the back pressure in the line, as the pressure increases are based on the restrictor's resistance to flow.

3.3.2 Fixed restrictors

Fixed restrictors are almost universally used in capillary SFC because of the low dead volumes required of capillary columns of i.d. 50–100 μm (Guthrie and Schwartz, 1986). Here the column is the determining factor of sample loading. However, with off-line SFE the sample loading should be determined by the size of the extraction cell and not by the restriction device. With the simplest fixed restrictor, e.g. a linear piece of fused silica 3–15 cm in length and 5–10 μm i.d. the sample loading is minimal. As a result the potential for plugging of this type of restrictor is greater than those of variable restrictors. Plugging most frequently occurs at the end of the restrictor where the depressurization of the supercritical fluid causes the analytes to drop out of solution. Two methods have been adopted to compensate for this problem. The first is to heat the restrictor tubing to avoid the build up of any ice and to keep the analytes from solidifying in the restrictor line. However, this may introduce problems of sample degradation at elevated temperatures. The second method is to place the end of the restrictor in an appropriate solvent for the target analytes and let the depressurizing fluid bubble through the solvent (Hills *et al.*, 1991). This does not compensate for extracts that are not soluble in the solvent. There are several variations on the fixed restrictor which try to compensate for the plugging problem. The frit restrictor (Markides *et al.*, 1986) (i.d. 50–100 μm) is one method of maintaining the back pressure without having very narrow internal diameters. The frit, essentially a linear restrictor with a packed end, allows multiple fluid paths and a better heat transfer, thus reducing ice formation, although high-molecular-weight analytes are more likely to plug the restrictor. In addition, diaphragm laser-drilled restrictors have been used as an interface with mass spectrometry because of their low dead volumes (Raynie *et al.*, 1989). The diaphragm must be able to support the high pressures involved. A common problem with fixed restrictors is that they are not very robust and can be easily broken, especially when they are handled frequently. One method of reducing this problem is to enclose the restrictor in a protective casing (Dionex Corporation, 1228 Titan Way, PO Box 3603, Sunnyvale, CA 94088-3603, USA). Another approach is to make the restrictor from a metal tube pinched at the end. The pinched restrictor (Green and Bertsch, 1988) is probably the most robust of all fixed restrictors but it is not very common because it is very difficult to manufacture them reproducibly. Most fixed restrictors have a limited lifetime and the problem of reproducibility of flow rates must be recognized when quantitative extraction is required. The main advantages of the fixed restrictors are that they are cheap and generally easy to replace. Thus, for simultaneous extractions from the same fluid source, fixed restrictors are more practical. Their small dead volumes also give them an advantage over variable restrictors when on-line coupling of analytical methods is required.

3.3.3 Variable restrictors

As discussed previously, the single largest advantage of variable restrictors over fixed restrictors is that they allow the independent control of the mass flow rate over the back pressure in the line, by constantly altering the size of orifice. Continual adjustment of the orifice requires a solenoid valve operated by a pressure transducer. The solenoid controls the mechanical action of opening and closing the orifice. In the event of an underpressure the orifice will close and momentarily stop the flow. The result of an overpressure is for the orifice to open and increase the flow rate. The restrictor is generally heated to around the temperature of the extraction cell; thus the adiabatic expansion of the depressurized fluid will not result in ice formation. With a sudden expansion there is less likelihood of analyte depositing in the sample line. In the event of a blocked line an overpressure situation would arise and the variable restrictor's response would be to open the orifice. This, quite often, is enough to unblock the line.

There are different methods of producing the rapid adjustment of the orifice. At the present time there are two commercially available instruments for SFE which use this type of technology. Jasco manufacture a back-pressure regulator (BPR) as part of their modular SFC/SFE instrument, and Hewlett-Packard use a variable nozzle restrictor in their off-line SFE 7680A extractor. Figure 3.7 illustrates a cross-sectional view of the Jasco BPR. The BPR was developed essentially for use in packed-column SFC, one of the main criteria for this being that it had a small dead volume (<10 μl). The orifice is controlled by a needle being forced into a valve seat. The action is rapid and the vibrations produced may reduce the risk of the restrictor blocking. The end of the restrictor is encased in a heating block and the temperature is pre-set on a control unit which also regulates the electronics for the solenoid valve and pressure transducer. At the top of the BPR there is a gap-adjustment screw. This is adjusted when the back pressure is altered in the control unit and allows the needle to vibrate freely in the valve seat. Variable restrictors are reliable and very robust. Their lifetime is indefinite and therefore, unlike fixed retrictors, extraction-condition repeatability is guaranteed. However, because of their relatively high cost they become impractical to use in simultaneous extraction systems.

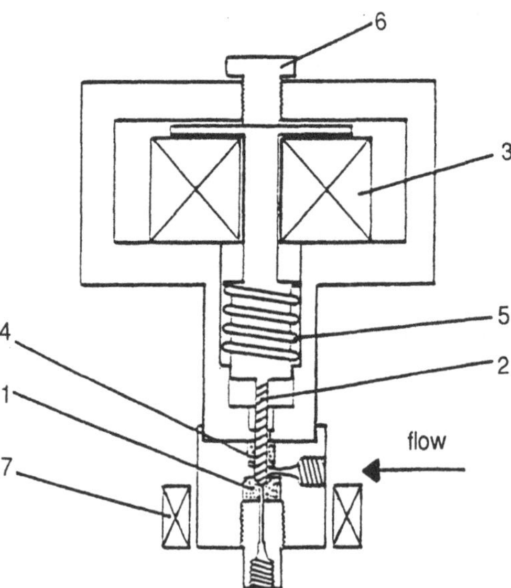

Figure 3.7 Back-pressure regulator: 1, valve seat; 2, valve needle; 3, needle-driven solenoid; 4, needle seal; 5, return spring; 6, gap-adjustment screw; 7, heater. Reproduced with permission from Saito *et al.* (1988).

3.4 Collection of extracts

Two methodologies have been adopted for collection of the extracted analytes after depressurization; these are on-line or off-line SFE. Generally, on-line SFE refers to the direct coupling of the analyte-containing supercritical fluid to a chromatographic separation system with appropriate detection. Off-line SFE allows for direct collection of the extracted analytes and retention for subsequent analysis.

3.4.1 On-line SFE

The connection of SFE directly with a chromatographic system has typically been done using SFC, GC or high performance liquid chromatography (HPLC). This has given rise to the production of dedicated instruments most notably using GC and SFC. Directly coupled GC is limited to volatile compounds while SFE–SFC can extend the analytical capability to compounds of higher molecular weight. The coupling of a chromatographic separation instrument with SFE frequently involves prior accumulation of the extract. This has been affected most typically using a heated (50–400 °C) transfer line, a cryogenic interface or a time-sequenced switching valve.

On-line SFE has the advantages of eliminating sample handling after loading in the extraction cell, thereby minimizing the risk of contamination. Also, all the extract is transferred for subsequent separation and detection. The major disadvantages are that the time for the entire analysis is approximately 1 h and the prerequisite to understand the nature of the extraction process prior to further analysis. This has the consequence that the whole instrument is not available for further analysis. However, the addition of a multi-hyphenated technique can allow unknown samples to be separated and spectroscopically investigated without the need for sample manipulation, using combinations such as supercritical fluid extraction–gas chromatography–Fourier transform infrared–mass spectrometry (SFE–GC–FTIR–MS).

3.4.1.1 Directly coupled SFE–GC. The on-line coupling of SFE with GC can be classified into two distinct categories (Lee and Markides, 1990): (i) those in which the method of extract accumulation is external to the GC; and (ii) those that utilize the GC for collection. These methods are summarized in Table 3.6.

3.4.1.2 Directly coupled SFE–SFC. In this case the extract is focused prior to SFC. Focusing can be achieved by several alternatives, including the following.

Table 3.6 Extract accumulation in directly coupled SFE–GC

Accumulation with or without GC	Accumulation method	References
External	Cold-trap	Liebman *et al.* (1989); Anderson *et al.*, (1989)
Internal	Retention gap at head of column	Wright *et al.* (1987, 1988); Onuska and Terry (1989)
Internal	On-column injection	Hawthorne and Miller (1986, 1987*a*); Wright *et al.* (1987, 1988); Hawthorne *et al.* (1988*a,b*, 1989*a,b*)
Internal	Split-splitless injection	Levy *et al.* (1987, 1989); Levy and Guzowski (1988); Levy and Rosselli (1989); Nielen *et al.* (1989); Hawthorne *et al.* (1990)
Internal	Switching valve	Wright *et al.* (1987); Levy *et al.* (1989); Onuska and Terry (1989)

(1) Dynamic focusing. This involves direct transfer of extracted analyte on to the column stationary phase. Transfer of extract is through uncoated or deactivated fused silica which prevents retention of analyte (Anton *et al.*, 1988; Hirata *et al.*, 1988; Ramsey *et al.*, 1989; Lee and Markides, 1990; Hedrick and Taylor, 1990; Raymer *et al.*, 1990), by a zero-dead-volume tee (Levy *et al.*, 1989) or by thermostatically controlled tubing (Engelhardt and Gross, 1988*a,b*).

(2) Cryogenic focusing. Extracted analyte is depressurized through a restrictor and collected on a cooled cryogenic trap accumulator (Anderson *et al.*, 1989; Xie *et al.*, 1989; Yocklovich *et al.*, 1989; Ashraf-Khorassani *et al.*, 1990; King, 1990). This has the distinct advantage of solvent peak removal from the chromatograph.

(3) Thermal modulator. A fused-silica open tubular column coated on the outside with an electrically conducting paint, which is rapidly heated using a pulsed electric current, is positioned at the head of the analytical column. The supercritical fluid containing the sample is introduced into the modulator. As the sample stream flows the analyte is adsorbed on the stationary phase. The analyte is desorbed as a discrete pulse by application of the electric current and transferred to the analytical column (Mitra and Wilson, 1990).

(4) Collection on a solid support. The extract is retained on a solid support material prior to desorption either by an increase in temperature or alternatively by passage of supercritical fluid (McNally and Wheeler, 1988*a*; Niessen *et al.*, 1988; Saito *et al.*, 1989).

(5) Static extraction. Extraction is done for a specified time period and the extract is retained in a sample loop. Switching of the injection valve allows the loop to be flushed with supercritical fluid mobile phase direct to the column (Sugiyama *et al.*, 1985; Jackson *et al.*, 1986; Skelton *et al.*, 1986; Engelhardt and Gargus, 1988; Jahn and Wenclawiak, 1988; McNally and Wheeler, 1988*b*; Raynor *et al.*, 1988; Thiebaut *et al.*, 1989; Wheeler and McNally, 1989).

3.4.1.3 Directly coupled SFE–HPLC. The ability of SFE to extract analytes of medium to low polarity allows for chromatographic separation using directly coupled GC and SFC. It is perhaps for this reason that directly coupled HPLC is rare (Hawthorne, 1990). This does not mean, however, that HPLC is never appropriate. The application of coupled SFE–HPLC is more specialized (Unger and Roumeliotis, 1983; Davies *et al.*, 1988; Nair and Huber, 1988; Subra and Boissinot, 1991) than the other combinations and more dependent upon the sample type.

An excellent review on the coupling of supercritical fluid extraction with chromatographic techniques has been provided by Vannoort *et al.* (1990).

3.4.1.4 Directly coupled SFE–MS. Chromatographic coupling of GC, HPLC and SFC to a mass spectrometer enables quantification and structural information to be obtained. The direct coupling of SFE to a chromatographic separation technique, as described above, obviously allows for the interpretation of complex samples by mass spectrometry. However, the direct coupling of SFE to a mass spectrometer has the potential benefits of minimal sample handling and preparation with no discrete extraction, purification or derivatization required, thus facilitating analysis and simplifying interpretation (Kalinoski *et al.*, 1986). The entire extract from the direct fluid injection interface was injected into the CI region of a dual EI–CI source and analysed using a triple-quadrupole mass spectrometer.

3.4.2 Off-line SFE

In contrast to on-line SFE, the capability to extract analytes off-line allows for a flexible approach to be used for sample separation and/or detection. The benefits of off-line SFE are summarized in Table 3.7.

The extracted analytes can be collected by several methods. These include: passing the supercritical fluid through a column packed with chromatographic material (Figure 3.5) (Sugiyama *et al.*, 1985; Schneiderman *et al.*, 1987; Mulcahey *et al.*, 1991); bubbling the condensed fluid through a small amount of solvent (Hawthorne and Miller, 1987*b*; Alexandrou and Pawliszyn, 1989; Campbell *et al.*, 1989; Hawthorne *et al.*, 1989*b*; DeRoos and Bicking, 1990; Lopez-Avila *et al.*, 1990; Sandra *et al.*, 1990; Swanson and Richter, 1990; Furton and Rein, 1991*a*,*b*; Hawthorne *et al.*,

Table 3.7 Benefits of off-line SFE

(a) High sample throughput
(b) Consideration of sample preparation only
(c) Chromatograph not idle for long periods of time, i.e. 10 min–1 h
(d) No understanding of chromatographic separation conditions required
(e) Not sample-size limited, therefore more reproducible sampling
(f) Variety of detectors possible
(g) Extracted analyte available for multiple analyses
(h) Familiarity with extraction conditions for specific sample types
(i) Dynamic extraction done without any valves between extraction cell and collector, therefore reduced risk of analyte loss
(j) Static extraction ensures smaller supercritical fluid consumption and ability to extract from larger samples

1991; Ma *et al.*, 1991; Rein *et al.*, 1991; Subra and Boissinot, 1991); and allowing the supercritical fluid-containing sample to expand into an empty container with or without cryogenic cooling (Brady *et al.*, 1987; McNally and Wheeler, 1988*a*; Ndiomu and Simpson, 1988; King, 1989; Kassim and Hameed, 1990; Richards and Campbell, 1991). A study on the reproducibility of GC injections following off-line SFE and collection of analytes in solvent was described by Swanson and Richter (1990). It was believed that during the extraction process the solvent for analyte collection is cooled sufficiently to allow saturation by the expanding carbon dioxide to occur. The workers reported that a significant deterioration in reproducibility occurred as a result. The results were compared with samples that were degassed by sonication for 1 min. Similar results were also obtained by sonicating samples prior to SFC.

The ability to monitor the progress of an extraction *in situ* with respect to time provides valuable information indicating the completeness of the extraction and/or that the maximum amount of analyte has been dissolved. Obviously, information pertaining to the kinetic status of the extraction is also inherent in the observed signal. Several groups have applied an in-line monitor to observe extraction phenomena. Typically, the in-line monitor used is an ultraviolet (UV)–visible spectrophotometer (Sugiyama *et al.*, 1985). This facilitates the detection of analytes containing a chromophore. Also, the detection is aided by the fact that carbon dioxide is transparent to about 190 nm. The use of other spectroscopic techniques, such as infrared spectroscopy has been hindered, despite their ability to provide information relating to functional groups, because carbon dioxide gives rise to strong molecular absorptions which totally obscure significant portions of the infrared (IR) region. This has been one of the reasons why researchers have investigated the use of alternative supercritical fluids. Supercritical xenon with a critical temperature of 16.8 °C and pressure of 58.0 atm has been recognized as the most suitable supercritical fluid when used in combination with FTIR. This combination has principally been applied to the detection of chromatographic eluents in SFC (Healy, 1991). Subra and Boissinot (1991) used the commercially available sample preparation accessory fitted with an in-line UV–visible detector to investigate extraction from brown algae. These workers used the in-line monitor to select the extraction time for the sample. By waiting until a constant UV-detector response was achieved they were able to achieve maximum recovery of extract from the algae by using a range of extraction pressures and temperatures. Alternatively, Sandra *et al.* (1990) were able to monitor the extent of extraction by removing 1 μl aliquots with a syringe at selected time intervals. Subsequent analysis of the extract can be monitored using capillary GC or SFC. This method is not ideal, as it does not provide immediate information on the extraction process. Also, subsequent analysis by a chromatographic separation technique increases the extraction time.

3.5 Method development

The key to reproducible and efficient SFE is in maintaining control over the parameters that effect that extraction (Table 3.2). Fortunately the technology of well-established chromatography techniques can easily be adapted to the requirements of SFE. High-pressure syringe and reciprocating pumps can deliver constant pressures at exact flow rates, while chromatographic ovens can rapidly achieve isocratic conditions to maintain precise supercritical temperature conditions. Exploited properly, the choice of extraction conditions can provide a degree of selectivity with much shorter time scales than traditional liquid–liquid or liquid–solid extraction.

3.5.1 Sample type

The first and most important part of any extraction is to determine the nature of the sample analyte–matrix combination. Is the analyte soluble in the extraction fluid, and if so can the analyte be extracted from the sample matrix? Of course, the answer to this is quite often unknown, but there are some simple rules which will help to establish whether an analyte can be extracted.

If a sample is to be extracted, then it must first demonstrate sufficient solubility in the supercritical medium. With carbon dioxide this is limited to low to moderately polar compounds. Hydrocarbons, alcohols, esters, aldehydes and ketones of relatively low polarity show a high solubility in supercritical carbon dioxide at relatively low densities (0.2–0.4 g ml^{-1}). The effect of substituting more polar groups such as a carbonyl or hydroxyl is to increase the polarity and increase the minimum density at which solubility is observed. Similarly, increasing molecular weight results in a drop in solubility of the analyte. Polymer samples such as poly(ethylene glycols) (PEGs), polystyrenes, polyethylenes and surfactants solubilize typically above 0.7 g ml^{-1}. Polar compounds such as amino acids and many pharmaceutical molecules show limited solubility in pure carbon dioxide. The way to handle such samples is either to use another, more polar, supercritical fluid and lose the favourable characteristics of carbon dioxide (Table 3.1), or to enhance the solvent strength by the addition of a modifier.

After it has been established that the analytes of interest can be solubilized in the supercritical fluid, the next step is to create enough mass transfer to transport analytes from the sample matrix. The rate of extraction is influenced by the diffusion rate of supercritical fluid through the sample matrix (Bartle *et al.*, 1991). For solid samples the particle size is critical to the diffusion rate. Freeze grinding and sieving are techniques that can be utilized to increase the surface area and hence improve diffusion through the sample matrix (King, 1989). For liquid samples, diffusion is a problem and hence it is desirable to suspend them on an inert solid phase such as Celite before extracting. Aqueous-based samples may be dried or pre-concentrated on to absorbent C_{18} extraction disks, provided that the analytes are not lost (Kane *et al.*, 1991). High-viscosity liquids such as creams and pastes can be homogenized with Celite by melting and then mixing. However, this runs the risk of sample degradation. An alternative method is to smear the sample on to a piece of glass. With environmental analysis of plant and animal material, particular problems can occur with low-melting-point fats and oils extruding through the supercritical fluid sample line. Restriction devices should be heated to avoid possible plugging while the sample cell should be maintained around the critical temperature (T_c) with these samples.

The main variables that are utilized in SFE are the supercritical fluid's density or pressure, temperature, the time the fluid passes through the sample, and (in the case of variable restriction) the flow rate. The correct choice of these variables can produce total extraction, while conversely an incorrect choice can fail to produce any extraction. The density or pressure variable is generally the most effective and should be optimized first. The threshold pressure is a term developed by Giddings *et al.* (1968) to describe the point where the analyte becomes soluble in the fluid. It is desirable to extract slightly above the threshold pressure to minimize the extraction of unwanted interferences. However, with interferences of similar characteristics to the analyte's (e.g. structure), the problem will still remain.

The temperature of the fluid should be elevated above the critical temperature to utilize supercritical conditions. Extraction recoveries are quite often improved by raising the temperature well above the critical temperature. This, at first, may seem strange as the consequences of increasing the temperature at a fixed pressure are to decrease the solvent density. At higher temperatures the rate of increase in the solvent's density is much reduced; e.g. at 100 °C a 200 bar increase in pressure may raise the density by 0.5 g ml^{-1} whereas over the same pressure increase at 40 °C the increase in density is nearer to 0.8 g ml^{-1}. However, the diffusion coefficient of the supercritical fluid increases with increasing temperature (as the fluid tends towards a more gas-like state) and as a result the mass transfer will increase, producing a faster extraction. King (1989) explained the improved extraction of triglycerides with increasing temperature above a pressure of 250 atm in terms of compatible solubility parameters of the analytes and supercritical fluid. The swelling of the sample matrix with increased temperature can be observed, especially with polymer-type samples (Shim and Johnston, 1989).

The relationship of time with the extraction efficiency is not, as one might expect, linear. As an extraction proceeds there is a drop in the amount extracted per unit time after approximately the first 50% of analyte is extracted. Figure 3.8 shows the relationship of extraction efficiency with time for a typical extraction. The last 20–30% of extraction is governed by diffusion processes. The use of elevated temperature and increased flow rates may improve the extraction efficiency for this last portion of the extraction.

When variable restrictors are used for SFE, there is independent control of the mass flow rate over the back pressure. An increase in the fluid flow results in an increase in the volume of fluid passed through the extraction cell and thus an increase in the diffusion. There are two flow modes in which extraction can be done, static and dynamic. Static flow must always be followed by a period of dynamic flow to ensure that there is enough mass transfer. The advantage of using static extraction is that in diffusion-controlled extraction there may be an enhanced recovery, particularly when larger cell volumes are used.

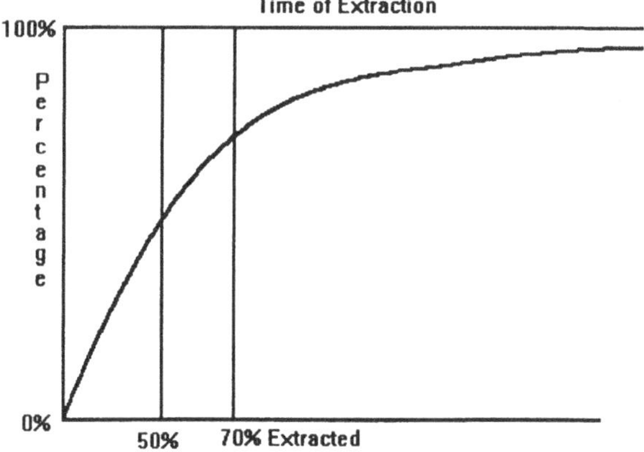

Figure 3.8 Percentage extraction versus time of extraction.

Table 3.8 Method development for supercritical fluid extraction

Sample type	Intermediate volatility	Involatile non-polar	Involatile non-polar, higher MW	Involatile polar
Pressure or density	Low	Intermediate	High	High with polar modifiers
Temperature	Low near critical temperature	Low near critical temperature	Increase temperature	Increase temperature >modifier boiling point
Time	Generally short but will vary on sample matrix	Short, longer with less soluble analytes	Increase time as solubility decreases	Increase time as solubility decreases
Flow rate	Low	Increase to improve diffusion	Increase to improve diffusion	Low, allow interaction of modifier

The potential for class-selective extraction in SFE is a major attraction of the technique. Selective extraction is controlled by the solvent strength and hence the supercritical fluid's density. If there is sufficient difference between the target analytes' solubilities, a stepwise increase in the supercritical density will allow selective extraction to be achieved. However, this is wholly dependent on the analytes' solubilities, and as a result selective extraction of a series of similar compounds is not possible. The solvating properties of supercritical fluids have also made them useful for fractionation and purification purposes (Yilgor and McGrath, 1984).

Table 3.8 summarizes an approach that may be taken for a particular sample type. It is not definitive and is useful only as a guide. At the present time optimization of extraction conditions is highly empirical. With increasing knowledge of the interactions that govern the extraction process the choice of conditions will become more precise. However, the large number of analyte–matrix combinations means that unique sample types are quite often encountered in SFE.

3.6 Conclusions

The versatility and flexibility of analytical SFE have been shown. However, most instrumentation is designed to accommodate only carbon dioxide as the supercritical fluid. This does present some shortcomings with respect to the extraction of more polar analytes. Instrumental development on the addition of modifiers is slow in development, relying on premixed cylinders, the use of a second piston pump, addition to syringe pump or spiking of sample. There is a requirement for a standard method of providing reproducible modifier additions. This would then allow the advancement of analytical SFE using carbon dioxide.

References

Alexandrou, N. and Pawliszyn, J., (1989), Supercritical fluid extraction for the rapid determination of polychlorinated dibenzo-*p*-dioxins and dibenzofurans in municipal incinerator fly ash. *Anal. Chem.*, **61**, 2770–2776.

Anderson, M.R., Swanson, J.T., Porter, N.L. and Richter, B.E., (1989), Supercritical fluid extraction as a sample introduction method for chromatography. *J. Chromatogr. Sci.*, **27**, 371–377.

Anton, K., Menes, R. and Widmer, H.M., (1988), Direct coupling of CO_2 fluid extraction with capillary supercritical fluid chromatography. *Chromatographia*, **26**, 221–223.

Ashraf-Khorassani, M., Kumar, M.L., Koebler, D.J. and Williams, G.P., (1990), Evaluation of coupled supercritical fluid extraction–cryogenic collection–supercritical fluid chromatography (SFE–CC–SFC) for quantitative and qualitative analysis. *J. Chromatogr. Sci.*, **28**, 599–604.

Bartle, K.D., Clifford, A.A., Hawthorne, S.B., Langenfeld, J.J., Miller, D.J. and Robinson, R., (1990), A model for dynamic extraction using a supercritical fluid. *J. Supercritical Fluids*, **3**, 143–149.

Bartle, K.D., Boddington, T., Clifford, A.A., Cotton, N.J. and Dowle, C.J., (1991), Supercritical fluid extraction and chromatography for the determination of oligomers in poly(ethylene terephthalate) films. *Anal. Chem.*, **63**, 2371–2377.

Brady, B.O., Kao, C.P.C., Dooley, K.M., Knopf, F.C. and Gambrell, R.P., (1987), Supercritical extraction of toxic organics from soils. *Ind. Eng. Chem. Res.*, **26**, 261–268.

Brunner, E., Hultenschmidt, W. and Schlichtharle, G., (1987), Fluid mixtures of high pressures. IV. Isothermal phase equilibria in binary mixtures consisting of (methanol + hydrogen or nitrogen or methane or carbon monoxide or carbon dioxide). *J. Chem. Thermodynamics*, **19**, 273–291.

Campbell, R.M., Meunier, D.M. and Cortes, H.J., (1989), Supercritical fluid extraction of chlorpyrifos methyl from wheat at part per billion levels. *J. Microcol. Sep.*, **1**, 302–208.

Davies, I.L., Raynor, M.W., Kithinji, J.P., Bartle, K.D., Williams, P.T. and Andrews, G.E., (1988), Instrumentation: SFE, LC, SFC–GC interfacing. *Anal. Chem.*, **60**, 683A–702A.

DeRoos, F.L. and Bicking, M.K.L., (1990), Supercritical fluid extraction for the determination of PCDDs and PCDFs in soil. *Chemosphere*, **20**, 1355–1361.

Engelhardt, H. and Gross, A., (1988*a*), On-line extraction and separation by supercritical fluid chromatography with packed columns, *J. High Resolut. Chromatogr., Chromatogr. Commun.*, **11**, 38–42.

Engelhardt, H. and Gross, A., (1988*b*), Extraction of pesticides from soil with supercritical CO_2. *J. High Resolut. Chromatogr., Chromatogr. Commun.*, **11**, 726.

Engelhardt, H. and Gross, A., (1991), Supercritical fluid extraction and chromatography: potential and limitations. *Trends Anal. Chem.*, **10**, 64–70.

Engelhardt, W.G. and Gargus, A.G., (1988), Supercritical fluid extraction for automated sample preparation. *American Laboratory*, **20**, 30–32.

Furton, K.G. and Rein, J., (1991*a*), The quantitative effect of microextractor cell geometry on the analytical supercritical fluid extraction efficiencies of environmentally important components. *Chromatographia*, **31**, 297–299.

Furton, K.G. and Rein, J., (1991*b*), Effect of microextractor cell geometry on supercritical fluid extraction recoveries and correlations with supercritical fluid chromatographic data. *Anal. Chim. Acta*, **248**, 263–270.

Giddings, J.C., Meyers, M.N., McLaren, L. and Keller, R.A., (1968), High pressure gas chromatography of nonvolatile species. *Science*, **162**, 67–73.

Gorner, T., Dellacherie, J. and Perrut, M., (1990), Comparison of helium head pressure carbon dioxide and pure carbon dioxide as mobile phases in supercritical fluid chromatography. *J. Chromatogr.*, **514**, 309–316.

Green, S. and Bertsch, W., (1988), Simple restrictors for capillary column supercritical fluid chromatography. *J. High Res. Chromatogr., Chromatogr. Commun.*, **11**, 414–415.

Guthrie, E.J. and Schwartz, H.E., (1986), Integral pressure restrictors for capillary supercritical fluid chromatography, *J. Chromatogr. Sci.*, **24**, 236–241.

Hawthorne, S.B., (1990), Analytical scale supercritical fluid extraction. *Anal. Chem.*, **62**, 633A–642A.

Hawthorne, S.B. and Miller, D.J., (1986), Extraction and recovery of organic pollutants from environmental solids and tenax-GC using supercritical carbon dioxide. *J. Chromatogr. Sci.*, **24**, 258–263.

Hawthorne, S.B. and Miller, D.J., (1987*a*), Directly coupled supercritical fluid extraction–gas chromatographic analysis of polycyclic aromatic hydrocarbons and polychlorinated biphenyls from environmental solids. *J. Chromatogr.* **403**, 63–76.

Hawthorne, S.B. and Miller, D.J., (1987*b*), Extraction and recovery of polycyclic aromatic hydrocarbons from environmental solids using supercritical fluids. *Anal. Chem.*, **59**, 1705–1708.

Hawthorne, S.B., Krieger, M.S. and Miller, D.J., (1988*a*), Analysis of flavor and fragrance compounds using supercritical fluid extraction coupled with gas chromatography. *Anal. Chem.*, **60**, 472–477.

Hawthorne, S.B., Miller, D.J. and Krieger, M.S., (1988*b*), Rapid extraction and analysis of organic compounds from solid samples using coupled supercritical fluid extraction/gas chromatography. *Fresenius Z. Anal. Chem.*, **330**, 211–215.

Hawthorne, S.B., Miller, D.J. and Krieger, M.S., (1989*a*), Coupled SFE–GC: A rapid and simple technique for extracting, identifying and quantitating organic analytes from solids and sorbent resins. *J. Chromatogr. Sci.*, **27**, 347–354.

Hawthorne, S.B., Krieger, M.S. and Miller, D.J., (1989*b*), Supercritical carbon dioxide extraction of polychlorinated biphenyls, polycyclic aromatic hydrocarbons, heteroatom-containing polycyclic aromatic hydrocarbons, and *n*-alkanes from polyurethane foam sorbents. *Anal. Chem.*, **61**, 736–740.

Hawthorne, S.B., Miller, D.J. and Langenfeld, J.J., (1990), Quantitative analysis using directly coupled supercritical fluid extraction–capillary gas chromatography (SFE–GC) with a conventional split/splitless injection port. *J. Chromatogr. Sci.*, **28**, 2–8.

Hawthorne, S.B., Miller, D.J., Walker, D.D., Whittington, D.E. and Moore, B.L., (1991), Quantitative extraction of linear alkylbenzenesulphonates using supercritical carbon dioxide and a simple device for adding modifiers. *J. Chromatogr.*, **541**, 185–194.

Healy, M.A., Jenkins, T.J. and Poliakoff, M., (1991), Infrared detection in supercritical fluid chromatography using xenon. *Trends Anal. Chem.*, **10**, 92–97.

Hedrick, J.L. and Taylor, L.T., (1990), Supercritical fluid extraction strategies of aqueous based matrices. *J High Res. Chromatogr. , Chromatogr. Commun.*, **13**, 312–316.

Hills, J.W., Hill, H.H., Jr. and Maeda, T., (1991), Simultaneous supercritical fluid derivatization and extraction. *Anal. Chem.*, **63**, 2152–2155.

Hirata, Y., Nakata, F. and Horihata, M., (1988), Direct sample injection in supercritical fluid chromatography with packed fused silica column. *J. High Resolut. Chromatogr., Chromatogr. Commun.*, **11**, 81–84.

Jackson, W.P., Markides, K.E. and Lee, M.L., (1986), Supercritical fluid injection of high-molecular-weight polycyclic aromatic compounds in capillary supercritical fluid chromatography. *J. High Resolut. Chromatogr., Chromatogr. Commun.*, **9**, 213–217.

Jahn, K.R. and Wenclawiak, B., (1988), Direct on-line coupling of small subcritical and supercritical fluid extractors with packed column supercritical fluid chromatography. *Chromatographia*, **26**, 345–350.

Kalinoski, H.T., Udseth, H.R., Wright, B.W. and Smith, R.D., (1986), Supercritical fluid extraction and direct fluid injection mass spectrometry for the determination of trichothecene mycotoxins in wheat samples. *Anal. Chem.*, **58**, 2421–2425.

Kane, M., Dean, J.R., Hitchen, S.M., Tranter, R.L. and Dowle, C.J., (1991), Analysis of surfactants using supercritical fluid chromatography and extraction. *European Symposium on Analytical Supercritical Fluid Chromatography and Extraction*, Wiesbaden, Germany, 4–5 December, p. 20.

Kane, M., Dean, J.R., Hitchen, S.M., Dowle, C.J. and Tranter, R.L., (1993), An experimental design approach for supercritical fluid extraction. *Anal. Chim. Acta*, **271**, 83–90.
Kassim, D.M. and Hameed, M.S., (1990), Direct extraction-separation of essential oils from citrus peels by supercritical carbon dioxide. *Separation Sci. Technol.*, **24**, 1427–1435.
King, J.W., (1989), Fundamentals and applications of supercritical fluid extraction in chromatographic science. *J. Chromatogr. Sci.*, **27**, 355–364.
King, J.W., (1990), Applications of capillary supercritical fluid chromatography–supercritical fluid extraction to natural products. *J. Chromatogr. Sci.*, **28**, 9–14.
Lee, M.L. and Markides, K.E., (1990), In *Analytical Supercritical Fluid Chromatography and Extraction*. Chromatography Conferences, Inc., Provo, Utah, USA, Chapter 5.
Levy, J.M. and Guzowski, J.P., (1988). Characterization of gasolines using on-line multidimensional supercritical fluid chromatography/capillary gas chromatography. *Fresenius Z. Anal. Chem.*, **330**, 207–210.
Levy, J.M. and Rosselli, A.C., (1989), Quantitative supercritical fluid extraction coupled to capillary gas chromatography. *Chromatographia*, **28**, 613–616.
Levy, J.M., Guzowski, J.P. and Huhak, W.E., (1987), On-line multidimensional supercritical fluid chromatography/capillary gas chromatography. *J. High Resolut. Chromatogr., Chromatogr. Commun.*, **10**, 337–341.
Levy, J.M., Cavalier, R.A., Bosch, T.N., Rynaski, A.F. and Huhak, W.E., (1989), Multidimensional supercritical fluid chromatography and supercritical fluid extraction. *J. Chromatogr. Sci.*, **27**, 341–346.
Li, S.F.Y., Ong, C.P., Lee, M.L. and Lee, H.K., (1990), Supercritical fluid extraction and chromatography of steroids with Freon-22. *J. Chromatogr.*, **515**, 515–520.
Liebman, S.A., Levy, E.J., Lurcott, S., O'Neill, S., Guthrie, J., Ryan, T. and Yocklovich, S., (1989), Integrated intelligent instruments: Supercritical fluid extraction, desorption, reaction and chromatography. *J. Chromatogr. Sci.*, **27**, 118–126.
Lopez-Avila, V., Dodhiwala, N.S. and Beckert, W.F., (1990), Supercritical fluid extraction and its application to environmental analysis. *J. Chromatogr. Sci.*, **28**, 468–476.
Ma, X., Yu, X., Zheng, Z. and Mao, J., (1991), Analytical supercritical fluid extraction of Chinese herbal medicines. *Chromatographia*, **32**, 40–44.
Majors, R.E., (1991*a*), An overview of sample preparation. *Liq. Chromatogr. Gas Chromatogr.*, **4**(2), 10–14.
Majors, R.E., (1991*b*), Supercritical fluid extraction–an introduction. *Liq. Chromatogr. Gas Chromatogr.*, **4**(3), 10–16.
Markides, K.E., Fields, S.M. and Lee, M.L., (1986), Capillary supercritical fluid chromatography of labile carboxylic acids. *J. Chromatogr. Sci.*, **24**, 254–257.
McNally, M.E.P. and Wheeler, J.R., (1988*a*), Supercritical fluid extraction coupled with supercritical fluid chromatogrpahy for the separation of sulphonylurea herbicides and their metabolites from complex matrices. *J. Chromatogr.*, **435**, 63–71.
McNally, M.E.P. and Wheeler, J.R., (1988*b*), Increasing extraction efficiency in supercritical fluid extraction from complex matrices. Predicting extraction efficiency of diuron and linuron in supercritical fluid extraction using supercritical fluid chromatographic retention. *J. Chromatogr.*, **447**, 53–63.
Mitra, S. and Wilson, N.K., (1990), Thermal modulation interface between supercritical fluid extraction and supercritical fluid chromatography. *J. Chromatogr. Sci.*, **28**, 182–185.
Mulcahey, L.J., Hedrick, J.L. and Taylor, L.T., (1991), Collection efficiency of various solid-phase traps for off-line supercritical fluid extraction. *Anal. Chem.*, **63**, 2225–2232.
Nair, J.B. and Huber, J.W., III, (1988), On-line supercritical sample-preparation accessory for chromatography. *Liq. Chromatogr. Gas Chromatogr.*, **6**, 1071–1073.
Ndiomu, D.P. and Simpson, C.F., (1988), Some applications of supercritical fluid extraction. *Anal. Chim. Acta*, **213**, 237–243.
Nielen, M.W.F., Sanderson, J.T., Frei, R.W. and Brinkman, U.A. Th., (1989), On-line system for supercritical fluid extraction and capillary gas chromatography with electron-capture detection. *J. Chromatogr.* **474**, 388–395.
Niessen, W.M.A., Bergers, P.J.M., Tjaden, U.R. and Van Der Greef, J., (1988), Phase-system switching as an on-line sample pretreatment in the bioanalysis of mitomycin C using supercritical fluid chromatography. *J. Chromatogr.*, **454**, 243–251.
Onuska, F.I. and Terry, K.A., (1989), Supercritical fluid extraction of PCBs in tandem with

high resolution gas chromatography in environmental analysis. *J. High Resolut. Chromatogr., Chromatogr. Commun.*, **12**, 527–531.

Pawliszyn, J., (1990), Inexpensive fluid delivery system for supercritical fluid extraction. *J. High Resolut. Chromatogr., Chromatogr. Commun.*, **13**, 199–202.

Ramsey, E.D., Perkins, J.R., Games, D.E. and Startin, J.R., (1989), Analysis of drug residues in tissue by combined supercritical fluid extraction–supercritical fluid chromatography–mass spectrometry–mass spectrometry. *J. Chromatogr.*, **464**, 353–364.

Raymer, J.H., Smith, C.S., Pellizzari, E.D. and Velez, G., (1990), Pyrolysis coupled with capillary supercritical fluid chromatography. *J. Liq. Chromatogr.*, **13**, 1261–1283.

Raynie, D.E., Markides, K.E., Lee, M.L. and Goates, S.R., (1989), Back-pressure regulated restrictor for flow control in capillary supercritical fluid chromatography. *Anal. Chem.*, **61**, 1178–1181.

Raynor, M.W., Davies, I.L., Clifford, A.A., Williams, A., Chalmers, J.W. and Cook, B.W., (1988), Supercritical fluid extraction/capillary supercritical fluid chromatography/Fourier transform infrared microspectrometry of polycyclic aromatic compounds in a coal tar pitch. *J. High Resolut. Chromatogr., Chromatogr. Commun.*, **11**, 766–775.

Rein, J., Cork, C.M. and Furton, K.G., (1991), Factors governing the analytical supercritical fluid extraction and supercritical fluid chromatographic retention of polycyclic aromatic hydrocarbons. *J. Chromatogr.*, **545**, 149–160.

Richards, M. and Campbell, R.M., (1991), Comparison of supercritical fluid extraction, Soxhlet and sonication methods for the determination of priority pollutants in soil. *Liq. Chromatogr. Gas Chromatogr.*, **4**(7), 36–36.

Saito, M., Yamauchi, Y., Inomata, K. and Kottkamp, W., (1989), Enrichment of tocopherols in wheat germ by directly coupled supercritical fluid extraction with semipreparative supercritical fluid chromatography. *J. Chromatogr. Sci.*, **27**, 79–85.

Sandra, P., David, F. and Stottmeister, E., (1990), Recovery studies by off-line SFE. *J. High Resolut. Chromatogr., Chromatogr. Commun.*, **13**, 284–286.

Schneiderman, M.A., Sharma, A.K. and Locke, D.C., (1987), Determination of anthraquinone in paper and wood using supercritical fluid extraction and high performance liquid chromatography with electrochemical detection. *J. Chromatogr.*, **409**, 343–353.

Schoenmakers, P.J., (1991), Effects of modifiers in SFC. Symposium on Extraction and Chromatography with Supercritical Fluids, University of Keele, UK, 3–4 October.

Shim, J.J. and Johnston, K.P., (1989), Adjustable solute distribution between polymers and supercritical fluids. *AIChE J.*, **35**, 1097–1106.

Skelton, R.J., Jr., Johnson, C.C. and Taylor, L.T., (1986), Sampling considerations in supercritical fluid chromatography. *Chromatographia*, **21**, 3–8.

Subra, P. and Boissinot, P., (1991), Supercritical fluid extraction from a brown alga by stagewise pressure increase. *J. Chromatogr.*, **543**, 413–424.

Sugiyama, K., Saito, M., Hondo, T. and Senda, M., (1985), New double-stage separation analysis method. Directly coupled laboratory-scale supercritical fluid extraction–supercritical fluid chromatography, monitored with a multiwavelength ultraviolet detector. *J. Chromatogr.*, **332**, 107–116.

Swanson, J.T. and Richter, B.E., (1990), Improving reproducibility of GC injections following off-line supercritical fluid extraction. *J. High Resolut. Chromatogr., Chromatogr. Commun.*, **13**, 385–386.

Thiebaut, D., Chervet, J.P., Vannoort, R.W., de Jong, G.J., Brinkman, U.A. Th. and Frei, R.W., (1989), Supercritical fluid extraction of aqueous samples and on-line coupling to supercritical fluid chromatography. *J. Chromatogr.*, **477**, 151–159.

Unger, K.K. and Roumeliotis, P., (1983), On-line high pressure extraction high performance liquid chromatography. I. Equipment design and operation variables. *J. Chromatogr.*, **282**, 519–526.

Vannoort, R.W., Chervet, J.P., Lingeman, H., de Jong, G.J. and Brinkman, U.A. Th., (1990), Coupling of supercritical fluid extraction with chromatographic techniques. *J. Chromatogr.*, **505**, 45–77.

Wheeler, J.R. and McNalley, M.E., (1989), Supercritical fluid extraction and chromatography of representative agricultural products with capillary and microbore columns. *J. Chromatogr. Sci.*, **27**, 534–539.

Wright, B.W., Frye, S.R., McMinn, D.G. and Smith, R.D., (1987), On-line supercritical fluid extraction–capillary gas chromatography. *Anal. Chem.*, **59**, 640–644.

Wright, B.W., Fulton, J.L., Kopriva, A.J. and Smith, R.D., (1988), Analytical supercritical fluid extraction methodologies. Chapter 3. In *Supercritical Fluid Extraction and Chromatography: Techniques and Applications,* ACS Symposium Series No. 366, American Chemical Society, Washington, DC, USA.

Xie, Q.L., Markides, K.E. and Lee, M.L., (1989), Supercritical fluid extraction-supercritical fluid chromatography with fraction collection for sensitive analytes. *J. Chromatogr. Sci.*, **27**, 365–370.

Yilgor, I. and McGrath, J.E., (1984), Novel supercritical fluid techniques for polymer fractionation and purification. *Polymer Bulletin*, **12**, 491–497.

Yocklovich, S.G., Sarner, S.F. and Levy, E.J., (1989), A process application of supercritical fluid extraction and chromatography. *American Laboratory*, **5**, 26–32.

4 Supercritical fluid chromatography and extraction of pharmaceuticals

I.D. WILSON, P. DAVIS and R.J. RUANE

4.1 Introduction

Over the last decade high performance liquid chromatography (HPLC) has become firmly established as the pre-eminent separation technique for the chromatography and analysis of drugs and pharmaceuticals. Clearly it is against this background that the adoption of new techniques such as supercritical fluid chromatography (SFC) must be considered. The immediate prerequisite for any new chromatographic technique, such as SFC, seeking to establish itself in such an environment is the identification and demonstration of significant advantages over the existing methods. It also follows that the availability of reliable and efficient equipment is also an essential requirement before any technique can become routinely employed.

The perceived advantages that SFC would appear to bring to pharmaceutical analyses are: firstly, that SFC offers the possibility of performing liquid chromatography using detectors previously restricted to gas chromatography (GC) such as electron capture and flame ionisation detectors (ECD, FID); and secondly, that the interfacing of SFC with the mass spectrometer would appear to be less fraught with problems than is currently the case with HPLC. A further advantage of SFC compared to HPLC is that it offers a different range of chromatographic possibilities and selectivities. In particular SFC offers the possibility of performing reliable 'normal phase' (adsorption) chromatography, which still remains difficult in HPLC for relatively polar molecules such as drugs. Other advantages that have been claimed for SFC include both faster and more efficient chromatography than for the equivalent separation by HPLC and faster method development.

However, certainly in the field of pharmaceutical/bioanalysis SFC is still in its infancy, and it is by no means clear that it will ever be more than a niche technique. Here we have attempted to review briefly the current status of SFC in this area, and combined this overview with our own experiences of the technique. We have not attempted to provide an exhaustive review of this subject but have aimed to provide illustrative examples which we hope demonstrate the potential of SFC for the future.

Throughout the text the term SFC has been used generically to denote both supercritical and subcritical conditions. This has been done on the pragmatic basis that whilst such distinctions may be of academic interest it is the use of CO_2 as a mobile phase, however modified, and the quality of the separation which is of practical importance here.

4.2 SFC of pharmaceuticals

The following sections represent an attempt to review the current literature on the SFC of pharmaceuticals on both capillary and packed columns. This has been undertaken with the aim of providing information on both the types of compound that have been investigated and the conditions that have been employed for their SFC. In general only studies that provided sufficiently detailed descriptions of the methodology, and which therefore offer the possibility of their separations being successfully repeated elsewhere, have been discussed.

4.2.1 Alkaloids

Historically the alkaloids have provided a rich source of pharmaceutical products and many are still important today. SFC has been carried out on this class of compounds by two groups, Berry *et al.* (1986), who investigated the ergot alkaloids in extracts of *Claviceps purpurea*, and Janicot *et al.* (1988), on poppy alkaloids.

The SFC of the ergot alkaloids of both the clavine (agroclavine, festuclavine, elymoclavine, noragroclavine, chanoclavine I and II, norchanoclavine) and the peptide type (ergocryptine and bromocryptine mesilate) was demonstrated with ultraviolet (UV) (280 nm) and mass spectrometry (MS) for detection. Packed-column SFC was performed on the clavine-type alkaloids using an aminopropyl-bonded stationary phase (Spherisorb, 5 μm, 100 mm × 4.6 mm i.d.) and CO_2 modified with methanol as mobile phase. The conditions employed involved the use of CO_2 containing 10% methanol at 3 cm^3 min^{-1}, 75 °C and 365 bar for 2.5 min. The flow rate was then increased to 5 cm^3 min^{-1} and the methanol content raised to 15% at 2.8 min and then 20% after 5 min. This resulted in the last compound to be eluted (norchanoclavine II) having a retention time of about 10 min. Silica was also tested as a stationary phase but proved to be unsuitable for the SFC of these compounds. Ergocryptine and bromocryptine mesilate were also chromatographed on aminopropyl silica (APS) with CO_2 modified with 29% methoxyethanol at a flow rate of 4 cm^3 min^{-1}, 380 bar and 75 °C. The use of supercritical fluid chromatography–electron impact mass spectrometry (SFC–EIMS) for the analysis of a complex extract of *Claviceps purpurea* was also demonstrated.

The use of packed column SFC with UV (280 nm) and diode-array UV detection for the analysis of seven opium alkaloids (codeine, morphine, cryptopine, narcotine, papaverine, thebaine and ethylmorphine) (see also section 4.3) was described in some detail by Janicot *et al.* (1988). Separations were conducted on either silica (5 μm Lichrosorb Si 60) or aminopropyl-bonded silica (10 μm Lichrosorb-NH_2, 3 μm Spherisorb-NH_2) packed in 120 mm or 230 mm × 4.6 mm i.d. stainless-steel columns. The influence of the methanol content of the mobile phase, water content and the use of various amine modifiers was studied. As would be expected with increasing methanol content the retention of the alkaloids was reduced. With about 16% (w/w) of methanol in the mobile phase, retention time for morphine (the most retained analyte) was approximately 5 min on aminopropyl silica and 15 min on silica. Addition of amines (methyl-, ethyl-, and triethylamine), with SFC on silica, also led to a general decrease in retention and small changes in selectivity. On aminopropyl-bonded silica, increasing the amount of amine in the mobile phase resulted in a general increase in retention. The effect of water in the mobile phase was studied in both the presence and the absence of an amine modifier. In the case of silica, in the absence of an amine, a decrease in retention was noted with increasing water content (e.g. k' for thebaine was 45 at 0.715% and 36 at 1.42% water, w/w). Under similar conditions with the aminopropyl silica phase hardly any effect was noted. With an amine in the mobile phase the addition of water led to an increase in retention on the aminopropyl silica. For silica the authors stated that, in the presence of an amine additive, the water content must be kept as low as possible. On the basis of these investigations the best mobile phase for the analysis of mixtures of these alkaloids on silica was determined to be CO_2–methanol–methylamine–water 83.37:16.25:0.15:0.23 (w/w). For aminopropyl silica the equivalent mobile phase compositions were 82.95:16.25:0.50:0.30 (w/w) for the 10 μm Lichrosorb-NH_2 phase (120 mm × 4.6 mm i.d.) or 87.62:1.80:0.36:0.22 (w/w) with 3 μm Spherisorb-NH_2 (120 mm × 4.6 mm i.d.). These conditions gave analysis times of between 2 and 11 min depending upon the stationary phase employed (aminopropyl phases giving shorter analysis times). When the method was applied to a poppy straw extract with SFC on silica, four alkaloids were detected (papaverine, thebaine, codeine and morphine). This result was claimed as being similar to that obtained by HPLC, based on published data, but with a greatly reduced analysis time.

4.2.2 Amphetamines

The packed-column SFC of five amphetamines (amphetamine, methylamphetamine, phenethylamine, ephedrine and norephedrine) with UV detection at 269 nm was investigated by Veuthig and Haerdi (1990). SFC was performed following derivatisation with 9-fluorenmethyl chloroformate (FMOC-Cl). The reagent, which reacts with primary and secondary amines, was used to form non-polar, UV-absorbing, derivatives. The reaction was performed by mixing the sample in buffer at pH 9.5 (below pH 9.0 derivatisation is incomplete) with the reagent in acetone and leaving to stand for 10 min. The reaction mixture was then extracted with dichloromethane and the organic layer taken for chromatography. The stationary phases investigated were Hypersil ODS (10 μm) and Hypersil aminopropyl silica (APS) (5 μm, 300 mm × 3.9 mm i.d.) and Nucleosil-100 silica (5 μm, 200 mm × 4 mm i.d.) packed in stainless-steel columns. Of itself CO_2 was unable to elute the derivatives on any of these phases, and a polar modifier was required. Methanol, 2-propanol and acetonitrile were all investigated, with methanol proving to have the highest eluotropic strength. Thus a methanol concentration of 2.4% (v/v) was sufficient to elute all five analytes from the silica column in under 5 min (40 °C, 4 ml min^{-1}, 200 bar). A lower eluotropic strength was shown by 2-propanol under the same conditions, and methylamphetamine and amphetamine were not resolved. In the case of acetonitrile, poor results were obtained with broad peaks, long retention times and poor efficiencies. The authors concluded that the best separations were obtained on silica gel or aminopropyl-bonded silica gel, whilst the ODS material gave only short retention times. Different selectivities were noted between the silica and APS materials, with superior separation of ephedrine and norephedrine on the latter. The final chromatographic conditions used were CO_2-methanol (4.8%) at 4 ml min^{-1}, 40 °C and 200 bar. These conditions provided an analysis time of less than 4 min combined with good peak shape.

4.2.3 Avermectins

The avermectins form a group of potent antiparasitic agents comprising α-L-oleandrosyl-α-oleandroside derivatives of pentacyclic 16-membered lactones structurally related to the milbemycins. Despite their polarity and high molecular weight, packed-column SFC of these compounds has proved to be possible with both UV (238 nm) and MS detection (Lane, 1988). Chromatography was achieved on silica gel (Rainin Microsorb, 5 μm, 150 mm × 4.6 mm i.d.) using CO_2-methoxyethanol (93:7) with a solvent flow rate of 3 ml min^{-1} at 65 °C and 3100 psi. Under these conditions, with UV detection, 22,23-dihydroavermectin B_{1a}-aglycone, monosaccharide and disaccharide were separated in just under 9 min. Slightly modified conditions (5% methoxyethanol at 2750 psi) were used for on-line mass spectrometry resulting in slightly longer retention times. Mass spectra for all three compounds were obtained, using negative-ion CI (with ammonia as reagent gas), which provided both molecular-weight information and diagnostic fragmentation. The SFC–EIMS of the compounds resulted in diagnostic ions for the 16-membered lactone but only weak molecular weight data.

4.2.4 *Barbiturates*

The packed-column SFC of a total of nine barbiturates (barbitone, butobarbitone, amylobarbitone, pentobarbitone, talbutal, quinalbarbitone, methohexitone, phenobarbitone and heptobarbitone) was investigated in two studies by Smith and Sagani (1988, 1989*a*) with flame ionisation (Smith and Sagani, 1988) and UV (254 nm) (Smith and Sagani, 1989*a*) detection. Chromatography was performed on columns packed with either Spherisorb ODS-2 (5 μm, 200 mm × 3 mm i.d.) or polystyrene–divinylbenzene (PS–DVB) (5 μm, 150 mm × 4.6 mm i.d.) with either CO_2 alone, with FID, or modified with methanol (0–14.6%, w/w), for SFC with UV detection. In the absence of methanol no elution was observed from the ODS phase over the range of operating pressures examined (up to 240 kg cm^{-1} at 60 °C, flow rate not reported). However, in contrast on the PS–DVB phase all the analytes except phenobarbitone and heptabarbitone were eluted with a column pressure of 155 kg cm^{-1}. All the barbiturates eluted with a column pressure of 187 kg cm^{-1}. The capacity factors obtained under these conditions are given in Table 4.1. In the absence of an organic modifier, however, peak tailing and adsorption remained significant. In a subsequent study, therefore, the authors investigated the use of methanol as a modifier for SFC on either Ultrasphere ODS (5 μm, 250 mm × 4.6 mm i.d.) or the PS–DVB phase with UV rather than FID detection (precluded by the presence of the organic modifier). The addition of methanol (4.0%, 9.1%, and 14.6%, w/w, 2200 psi, 650 °C, flow rate not reported) to the CO_2 resulted in a sharpening of the peaks eluting from the

Table 4.1 Capacity factors of baribiturates following packed-column SFC on a PS-DVB phase with CO_2: effect of pressure (Smith and Sanagi, 1988)

	Capacity factor, k'	
Column pressure	155 kg cm^{-2}	187 kg cm^{-2}
Amylobarbitone	6.55	3.57
Barbitone	5.32	3.17
Butobarbitone	6.51	3.70
Heptabarbitone	—	13.69
Methohexitone	6.16	3.43
Pentobarbitone	6.56	3.78
Phenobarbitone	—	13.28
Quinalbarbitone	8.34	4.54
Talbutal	7.50	4.30

PS–DVB column and reduced retention. Relative retentions were observed to be very dependent on the proportion of organic modifier used. The capacity factors for these barbiturates obtained on the PS–DVB phase are given in Table 4.2.

Similar studies on the ODS phase with 4.2% and 8.4% (w/w) methanol (1950 psi, 60 °C) enabled good peak shapes to be obtained for most of the analytes, with retention decreasing with increasing methanol content of the mobile phase. The elution orders for the compounds were similar except for amylobarbitone and phenobarbitone. Capacity factors for the barbiturates on the ODS column are given in Table 4.3.

The use of open-tubular capillary SFC for the analysis of the anti-epileptic drug phenobarbitone in serum samples, with barbitone as an internal standard, was described by Wong and Dellafera (1990). Chromatography was performed on a 10 m SB-methyl-10 column of 50 μm i.d. A 5 m

Table 4.2 Capacity factors of barbiturates following packed-column SFC on a PS-DVB phase with CO_2–methanol (Smith and Sanagi, 1989*a*)

	Capacity factor, k'			
Methanol (%, w/w)	0	4.0	9.1	14.6
Amylobarbitone	6.55	1.74	0.78	0.40
Barbitone	5.32	1.34	0.67	0.37
Butobarbitone	6.51	1.73	0.78	0.39
Heptabarbitone	>24	5.07	1.98	0.98
Methohexitone	6.16	2.27	1.11	0.69
Pentobarbitone	6.56	1.75	0.79	0.39
Phenobarbitone	>24	4.87	1.91	0.91
Quinalbarbitone	8.34	2.19	0.95	0.48
Talbutal	7.50	2.06	0.92	0.47

Table 4.3 Capacity factors of baribiturates following packed-column SFC on C_{18}-bonded silica[a] with CO_2–methanol (Smith and Sanagi, 1989*a*)

	Capacity factor, k'	
Methanol (%, w/w)	4.2	8.4
Amylobarbitone	0.42	0.17
Barbitone	0.30	0.16
Butobarbitone	0.37	0.17
Heptabarbitone	0.78	0.35
Methohexitone	—	—
Pentobarbitone	—	—
Phenobarbitone	0.65	0.28
Quinalbarbitone	0.46	0.24
Talbutal	0.41	0.22

[a]Ultrasphere ODS.

fused-silica column (also 50 μm i.d.) was used to provide a retention gap. With CO_2 as mobile phase, a temperature of 120 °C and a density gradient from 0.25 to 0.6 g ml^{-1} (0.02 g ml^{-1} min^{-1}), phenobarbitone was eluted with a retention time of 6.8 min and barbitone at 5.6 min. Pressure programming from 100 to 300 atm (20 atm min^{-1}) at 120 °C gave retention times of 6.3 min and 7.3 min for barbital and phenobarbital, respectively. Preliminary results were also provided for a range of other anti-epileptics (phenytoin, secobarbitone and pentobarbitone).

The use of solid-phase extraction to prepare the samples, rather than liquid–liquid extraction with methylene chloride, was found to be necessary to provide adequate column life. The assay was shown to be comparable to a clinically established fluorescence polarisation assay.

4.2.5 Benzodiazepines

The separation of a total of eleven benzodiazepines has been investigated using packed-column SFC with UV detection at 254 nm (Smith and Sanagi, 1989*b*). The compounds investigated were chlorodiazepoxide, cloxazolam, diazepam, estazolam, ketazolam, loprazolam, lorazepam, lormetazepam, nordazepam, temazepam and triazolam. The chromatography of the test compounds was studied on either Ultrasphere ODS or cyano-bonded silica gel (5 μm, 250 mm × 4.6 mm i.d.) or polystyrene–divinylbenzene (PS–DVB) phases (150 mm × 4.6 mm i.d.) with CO_2-methanol eluents. Modifiers were necessary because of the total retention of all of the test compounds on ODS silica, an effect ascribed to the presence of free silanols. Similarly, no elution was observed from the cyano column with CO_2 alone as mobile phase. On the PS–DVB column the bulk of the analytes were retained with only lorazepam, temazepam and diazepam eluted (as broad late-running peaks) with CO_2 as the mobile phase. Addition of either methanol or acetonitrile was observed to reduce retention times and improve peak shape considerably. The use of 4.3% methanol (w/w) (at 60 °C and 2515 psi, the flow rate was not reported) allowed the elution of all the benzodiazepines except loprazolam. Further increasing the methanol content to 9.7% and 15.3% (w/w) gave the expected decrease in retention, with loprazolam eluting in under 12 min at the higher concentration. Capacity factors for these compounds on the PS–DVB column are given in Table 4.4.

On the ODS phase, increasing the methanol content of the mobile phase from 4% to 16.4% (w/w, at 60 °C and 2470 psi) resulted in a marked decrease in the capacity factors obtained for these benzodiazepines. As shown in Table 4.5, for the same proportion of methanol in the solvent the ODS phase was much less retentive than the PS–DVB material. It was also noted that the peak shapes obtained for three hydroxylated compounds, lormatazepam, lorazepam and temazepam, were subject to considerable

Table 4.4 Capacity factors of benzodiazepines following packed-column SFC on a polystyrene–divinyl benzene[a] phase with CO_2–methanol (Smith and Sanagi, 1989*b*)

	Capacity factor, *k'*		
Methanol (%, w/w)	4.3	9.7	15.3
Chlorodiazepoxide	18.45	4.81	2.39
Cloxazolam	24.70	6.70	3.03
Diazepam	18.68	6.11	2.80
Estazolam	32.82	6.83	2.23
Ketazolam	20.44	6.22	2.25
Loprazolam	—	26.24	7.67
Lorazepam	16.48	4.10	1.40
Lormetazepam	16.24	5.57	2.20
Nordazepam	15.61	4.46	1.82
Temazepam	16.84	5.62	2.28
Triazolam	28.57	5.78	1.94

[a]PLRP-S.

tailing at low proportions of methanol which improved as the methanol content was increased. This phenomenon was attributed to silanophilic interactions. In comparison with the reversed-phase HPLC of the analytes on the same phase, using a methanol–phosphate buffer eluent (pH 7.25, 55:45) it was noted that there was essentially a reversal in elution order, i.e. a 'normal phase' type of chromatography. For both ODS and PS–DVB phases the selectivity of the separation was shown to be highly dependent on the organic modifier content of the CO_2.

The limited results obtained on the cyano-bonded silica (12.8% methanol w/w, 60 °C, 2470 psi) showed good peak shapes for all the analytes,

Table 4.5 Capacity factors of benzodiazepines following packed-column SFC on a C_{18} bonded phase[a] with CO_2–methanol (Smith and Sanagi, 1989*b*)

	Capacity factor, *k'*			
Methanol (%, w/w)	4.0	8.3	12.7	16.4
Chlorodiazepoxide	5.12	1.19	0.66	0.46
Cloxazolam	2.88	1.15	0.73	0.58
Diazepam	1.56	0.75	0.50	0.31
Estazolam	5.63	1.05	0.52	0.28
Ketazolam	1.54	0.76	0.51	0.31
Loprazolam	—	—	—	3.17
Lorazepam	7.91	1.44	0.61	0.35
Lormetazepam	5.15	1.39	0.79	0.44
Nordazepam	1.64	0.65	0.40	0.23
Temazepam	5.86	1.54	0.83	0.48
Triazolam	6.19	1.08	0.53	0.26

[a]Ultrasphere ODS.

Table 4.6 Capacity factors of benzodiazepines following packed-column SFC on a cyano-bonded phase[a] with CO_2–methanol (87.2:12.8, w/w) (Smith and Sanagi, 1989*b*)

	Capacity factor, k'
Chlorodiazepoxide	2.83
Cloxazolam	4.76
Diazepam	1.25
Estazolam	5.19
Ketazolam	1.30
Loprazolam	>19
Lorazepam	2.39
Lormetazepam	1.70
Nordazepam	1.63
Temazepam	1.58
Triazolam	6.42

[a]Ultrasphere CN.

including the hydroxyl-containing compounds (Table 4.6). Some similarities with the HPLC separation (hexane–isopropanol, 90:10) were noted, e.g. longer retentions for temazepam and lorazepam compared to diazepam and nordazepam, but chlorodiazepoxide was well retained on SFC but not on HPLC.

It was noted that the selectivities of the three phases studied were such that even with similar proportions of organic modifier the relative retentions of the benzodiazepines were very different.

4.2.6 Beta-blockers

A number of groups have reported results for the packed-column SFC of beta-blockers with UV or in-line radioactivity detection. As well as 'conventional' SFC with CO_2-methanol mobile phases, these studies have also included the use of ion-pairing reagents and chiral separations based on either the inclusion of a chiral counter-ion or the use of a chiral stationary phase. In our own studies we have obtained the best results by using aminopropyl-bonded silica gel columns (5 μm, 150 mm × 4.6 mm i.d.), where good chromatography, with relatively little peak tailing, was obtained with CO_2-methanol (9:1, v/v, 5 ml min^{-1}) at 50 °C and 3000 psi (Roberts and Wilson, 1990). Such conditions resulted in a retention time of about 5 min for propranolol. Typical k' values for a range of beta-blockers were: timalol, 2; atenolol, 10; betaxolol, 3; pindolol, 1; bupranolol, 2; pronethalol, 4; oxprenolol, 4. Labetolol was not eluted under these conditions. In a subsequent study (Ruane *et al.*, 1990), where [^{14}C]-propranolol was used to investigate the properties of a radio flow cell, 0.1% triethylamine was also included in the mobile phase to obtain the best possible peak

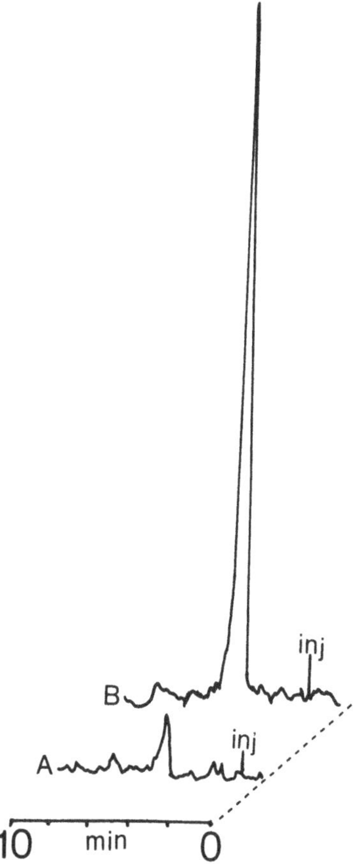

Figure 4.1 Packed-column SFC of [^{14}C]-propranolol on aminopropyl-bonded silica gel (5 μm, 150 mm × 4.6 mm i.d.) with CO_2–methanol (containing 0.1% triethylamine) (90:10, v/v) at 50 °C. (A) 10 000 dpm on column; (B) 100 000 dpm on column.

shape. Typical results are illustrated in Figure 4.1. In contrast to the results obtained on the aminopropyl phase on a cyano-bonded column (5 μm, 150 mm × 4.6 mm i.d.), no elution was observed with CO_2-methanol mixtures alone. However, the addition of triethylamine (TEA) (0.4%, v/v) did facilitate the elution of the beta-blocker with good peak shape. The SFC of propranolol on both amino and cyano columns is shown in Figure 4.2. Elution was not achieved, even in the presence of TEA, on alternative phases such as silica gel, alumina or nitro, whilst poor peak shape was obtained on both porous graphitic carbon (Hypercarb) and polymer-based columns (Rogel RP) (Roberts and Wilson, 1990).

An alternative approach, employing ion-pairing reagents, has also been explored by Steur *et al.* (1990) for the packed-column SFC of the beta-blockers pindolol, propranolol and bopindalol (as well as a number of other acidic and basic analytes). Chromatography was investigated using

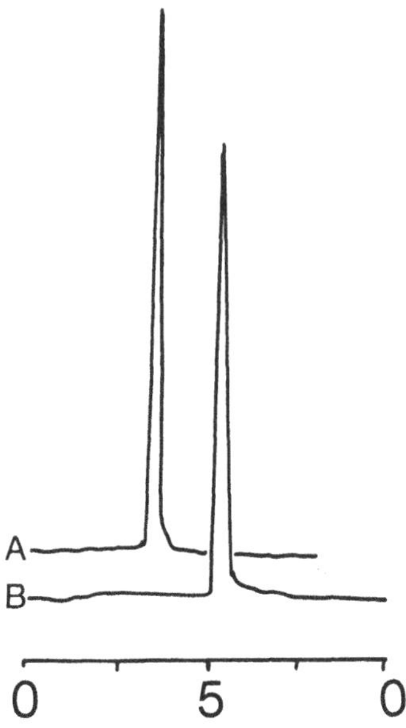

Figure 4.2 Packed-column SFC of propranolol on (A) cyanopropyl-bonded silica gel (150 mm × 4.6 mm i.d.) and (B) aminopropyl-bonded silica gel (150 mm × 4.6 mm i.d.) using CO_2–methanol (90:10 v/v). In the case of the cyanopropyl phase the addition of 0.4% TEA (v/v) to the methanol was required to obtain elution.

cyano (CS-MP Spheri 5, 100 mm × 4.6 mm i.d., Spherisorb CN, 100 mm × 4 mm i.d.), diol (OH-MP Spheri 10, 100 mm × 4.6 mm i.d.; LiChrospher 100, 250 mm × 4 mm i.d.) and silica gel (SS-MP Spheri 5) packing materials. As well as a variety of columns a range of ion-pair reagents was also tested, comprising: dimethyloctylamine, triethylamine, trioctylamine, tributylamine, tetrabutylammonium (as the bromide) and heptane sulphonic acid (as the sodium salt). For all the ion-pair reagents studied, 5% methanol or 10% acetonitrile was necessary in order that sufficient of the ion-pair reagent was dissolved in the mobile phase to ensure that a significant degree of ion-pairing occurred. In addition, it was necessary to maintain the density of the supercritical fluid above 0.5 g cm^{-3}. As expected, increasing the amount of ion-pair reagent in the eluent decreased retention, with a useful working range of concentrations from 1 to 4 mM recommended. Under these conditions most of the cationic test analytes eluted with good peak shape and short retention times on cyano phases, with diol phases also proving to be suitable. Thus for propranolol on the cyano phase (5 μm, 100 mm × 4.6 mm i.d.), 25 mM heptanesulphonic acid in methanol–CO_2 (20:80) at 60 °C and 225 bar gave a retention time of about 5 min. Addition of dimethyloctylamine (1.15 mM), as a competing base to counter the strong interactions of the basic analytes with the silica surface, under the same conditions resulted in a retention of about 4 min for propranolol. The effect of pressure on retention was also investigated, with k' values for bopindalol, pindolol and propranolol decreasing with increasing pressure. This study also illustrated the benefits that pressure programming can provide for the optimisation of SFC separations.

4.2.6.1 Chiral SFC of beta-blockers. The beta-blockers are usually administered as racemates. With the increasing awareness of the fact that the individual enantiomers present in a racemate can have quite different biologial effects, and be handled differently in biological systems, much effort has been expended on developing stereoselective assays. Chiral SFC has also been applied to the resolution of the enantiomers of the beta-blockers in a number of model systems. In the first example, *N*-benzoxycarbonylglycyl-L-proline (ZGP) was used to facilitate the separation of the beta-blockers pindalol, metoprolol, oxprenolol and propranolol, and DP1 201–106 as diasteriomeric ion-pairs (Steur *et al.*, 1988). Results for packed-column SFC with CO_2–acetonitrile as the mobile phase (under both super- and subcritical conditions) and a cyano-bonded phase (CS–GU, 100 mm × 4.6 mm i.d.) were described in some detail, whilst studies to investigate the suitability of additional phases (e.g. diol, phenyl and C_2) were alluded to. Because of the low solubility of ZGP in the mobile phase it was necessary to keep the density above 0.7 g cm^{-3} to prevent the precipitation of the reagent (the restrictor was continuously washed with acetonitrile for the same reason). TEA was also added to the mobile phase as a competing base to reduce peak tailing. No separation of the enantiomers was observed with concentrations of ZGP below 10 mM, but thereafter increasing the concentration of ZGP resulted in increased separation. Increasing the concentration of TEA resulted in slightly decreased separation. Increasing the pressure was associated with decreased retention but chiral selectivity was somewhat greater at lower pressure. Suitable supercritical fluid conditions for the separation of propranolol enantiomers were CO_2–acetonitrile (containing 35 mM ZGP and 5 mM TEA) 80:20 (v/v) at 60 °C and 225 bar. The corresponding conditions used for subcritical fluid chromatography were 20 °C and 250 bar. Subcritical fluid chromatography was found to be superior to both HPLC and SFC for the enantiomeric resolution of these compounds.

The direct separation of the enantiomers of a number of beta-blockers was achieved on a modified cellulose immobilised on to silica gel (Chiracel OD, 10 μm, 250 mm × 4.6 mm i.d.) by Lee *et al.* (1991) using both HPLC and SFC. The mobile phase used for SFC was CO_2 modified with methanol (80:20, v/v) at 35 °C and 200 bar, with a flow rate of 4 ml min^{-1}. Under these conditions the enantiomers of nadolol, betaxolol, pindolol, metoprolol, cicloprolol and propranolol were well resolved. A comparison with normal-phase HPLC on the same column showed that, whilst the separations were similar, SFC was faster and more efficient. Further studies on the separation of nadolol, which has two chiral centres and gave three peaks under these conditions, were also undertaken. Thus the first of the three eluted peaks, which had twice the peak area of the other two, was collected, derivatised with isopropyl isocyanate and subjected to further analysis by SFC on an amino column, on which it was resolved into two peaks. The SFC conditions employed for the amino column (Spherisorb, 5 μm, 100 mm × 4.6 mm i.d.) were CO_2–methanol (95:5, v/v), 25 °C and 200 bar at a flow rate of 3.0 ml min^{-1}.

More recently a detailed description of the chiral SFC of a total of ten beta-blockers (acebutalol, alprenolol, atenolol, betaxolol, metoprolol, oxprenolol, propranolol, beta-propranolol and pindolol) on two stationary phases derived from 3,5-dinitrobenzoyltyrosine has been reported by Siret *et al.* (1992). Packed-column (150 mm × 4.6 mm i.d.) separations were performed on (S)-ChyRoSine-A and an 'improved version' of this phase. The mobile phases involved in this work were based on CO_2 modified with methanol (4–10%) or methanol–dioxane (2:1, v/v) containing 1% of *n*-propylamine. The role of the amine appears to be the masking of residual silanols, as changing the proportion affected retention but not stereoselectivity. Similarly, the nature of the amine did not seem to be an important determinant of the stereoselectivity of the separation. Typical SFC conditions were an average pressure of 200 bar at 25 °C and 4 ml min^{-1}. Excellent separations were obtained with alpha values of between 1.11 and 1.71 depending upon the analyte. The results were both faster than and superior to the separation of the same analytes on these columns used in the normal-phase HPLC mode. Interestingly, when CO_2 was added to the normal-phase HPLC eluent (hexane–ethanol containing 1% of *n*-propylamine) at concentrations of up to 8% (v/v), retention was increased and stereoselectivity markedly improved. This led to the suggestion, supported by ^{1}H-NMR studies on propranolol, that the beta-blockers formed a complex with the CO_2 and the secondary amino and hydroxyl groups, leading to greater conformational rigidity and thus improved chiral discrimination.

4.2.7 Cephalosporins

Another class of antibiotics to be successfully chromatographed using packed-column SFC is the cephalosporins (Lane, 1988). Thus, packed-column SFC was performed on silica gel (Rainin Microsorb, 5 μm, 150 mm × 4.6 mm i.d.), with a mobile phase consisting of CO_2-methanol (95.5:4.5) and MS and UV detection (276 nm). The mobile phase was delivered at 4.5 ml min^{-1} at 3200 psi and 55 °C. These conditions facilitated the separation of the diastereoisomers of a cephalosporin ester (cefuroxime ester E47) in a little under 10 min, as well as the resolution of both compounds from a structural isomer. SFC–MS of these compounds gave characteristic electron impact (EI) spectra. The SFC system was also used to characterise impurities.

4.2.8 Cyclosporin

Cyclosporin is a cyclic undecapeptide widely used as an immunosuppressant in organ transplantation. With a molecular weight of 1202 it is at first sight an unpromising candidate for capillary SFC with unmodified CO_2. However, White *et al.* (1988) obtained good peak shape using a 9 m × 50 μm i.d. fused-silica capillary column coated with a 0.20 μm film of DB-5. The column was operated at 150 °C with an initial pressure of 2000 atm of CO_2. This pressure was maintained for 10 min and then a linear programme to 300 atm was employed at a rate of 10 atm min^{-1}. Under these conditions cyclosporin was eluted with a retention time of about 26 min. Detection was by FID; however, the authors pointed out that for clinical use SFC with nitrogen detection via the NPD would offer practical benefits.

4.2.9. Erythromycin A

Erythromycin A has been successfully chromatographed by packed-column SFC on amino-bonded silica gel (Spherisorb NH_2, 5 μm, 150 mm × 4.6 mm i.d.) with both UV (215 nm) and MS detection (Niessen *et al.*, 1989). The chromatographic conditions employed were CO_2-methanol (92:8) at 2.5 ml min^{-1}, 4000 psi and 65 °C. Under these conditions erythromycin A was eluted with a retention time of about 10 min. Mass spectra were obtained using positive CI which gave both molecular-weight and diagnostic fragmentation data. In addition minor impurities related to erythromycin A were also detected using SFC–MS by monitoring *m/z* 158.

4.2.10 Mefloquine

SFC with electron capture detection (ECD) was employed by Mount *et al.* (1990) for the analysis of the antimalarial drug mefloquine and an internal standard, D,L-erythro-α-(2)piperidyl)-2,8-bis(trifluoromethyl)-4-quinolinemethanol. Chromatography was performed on a 200 mm × 0.75 mm glass-lined steel column packed with 7 μm Zorbax BP silica gel, using supercritical pentane containing *n*-butylamine (0.15%) and methanol (1%) as mobile phase. *n*-Pentane was chosen rather than CO_2 based on the ease of addition of precise amounts of the modifiers to a liquid at room temperature and the reduced response of the ECD to *n*-pentane compared to CO_2. The column was conditioned for 4–6 h at 300 °C with the mobile phase at a pressure of 2100 kPa and for 10–20 min at the beginning of each working day. For analysis the column and transfer line were maintained at a temperature of 210 °C with the electron capture detector at 350 °C. Under these conditions retention times of 4 and 8 min were obtained for mefloquine and the internal standard respectively. Although this provided a sensitive method for the analysis of mefloquine, significant peak tailing was apparent.

4.2.11 Mitomycin C

The packed-column SFC of the anti-cancer drug mitomycin C following extraction from plasma, with UV detection at 360 nm, has been described by Niessen *et al.* (1988). Chromatography was performed on a Rosil C18 column (5 μm, 150 mm × 4.6 mm i.d.) with a precolumn (10 mm × 3.2 mm i.d.) prepared from the same material. When SFC on silica and aminopropyl phases was attempted, mitomycin C was not eluted from the column. Similar experiments on C_8 columns were also unsuccessful, resulting in poor peak shapes. On the C_{18} phase there was no elution with CO_2 alone as the mobile phase. However, elution was obtained following the addition of methanol to the mobile phase, and the conditions finally adopted used CO_2–methanol (88:12, v/v) at a flow rate of 2 ml min^{-1}. Chromatography was performed at a constant back pressure of 30 MPa and at a temperature of 50 °C (the compound is unstable at temperatures above 60 °C). Under these conditions retention times of about 2–3 min were obtained, and even with some peak tailing detection limits of 0.4 ng were reported.

The microbore packed-column SFC of mitomycin C was also reported by Musser and Callery (1990) on a 100 mm × 1 mm Nucleosil (3 μm) column with MS as the means of detection. The conditions used were 400 atm and 80 °C with CO_2 containing 5% methanol as mobile phase.

4.2.12 Non-steroidal anti-inflammatory drugs (NSAIDs)

The SFC of a variety of NSAIDs has been demonstrated on both capillary and packed-column systems.

Capillary SFC of ibuprofen, naproxen, flufenamic acid, phenylbutazone (see also later under drugs of abuse) and indomethacin (together with an unidentified impurity) has been shown by Lee *et al.* (1988) on a 10 m × 50 μm i.d. SB-methyl column. Supercritical CO_2 as the mobile phase at 1–3 ml min^{-1}, a column head pressure of 200 atm and a temperature of 150 °C were used. Detection was by 'charge exchange' (CE) MS, which gives 'pseudo' EI spectra (thus facilitating library searching), and representative spectra were obtained for these compounds at the 10–60 ng on-column level. The identification of the dicarboxylic acid metabolite of ibuprofen in an extract of horse urine was also demonstrated using an SB-octyl column.

More recently, Jagota and Stewart (1992*a*) have studied the capillary SFC of a range of NSAIDs (ibuprofen, flufenamic acid, fenoprofen,

Table 4.7 Capacity factors of non-steroidal anti-inflammatory drugs following capillary SFC on SB-biphenyl and SB-cyanopropyl columns with CO_2 (Jayota and Stewart, 1992*a*)

	Capacity factor, k'	
	SB-biphenyl	SB-cyanopropyl
Aspirin	—	1.4
Fenbufen	2.0	2.0
Fenoprofen	1.5	1.1
Ibuprofen	1.4	0.8
Indomethacin	2.6	—
Ketoprofen	2.0	1.5
Naproxen	1.7	1.4
Sulindac	3.3	1.0
Tolmetin	1.9	2.0

naproxen, tolmetin, fenbufen, ketoprofen, indomethacin, mefanamic acid, sulindac and aspirin) using three different stationary phases. SFC was performed isothermally at 130 °C using CO_2 as the mobile phase and flame ionisation detection. The columns investigated were a 5 m × 100 μm i.d. SB-methyl-100 (100% methylpolysiloxane), a 10 m × 50 μm i.d. SB-biphenyl-30 (30% biphenyl and 70% methylpolysiloxane) and a 10 m × 50 μm i.d. SB-cyanopropyl-50 (50% cyanopropyl and 50% methylpolysiloxane). The SB-methyl-100 column proved to be unsuited to the SFC of these compounds but good results were obtained with the other two phases. Compounds were eluted from these columns by using pressure programming. The programme used was as follows: 7 min at 100 atm, then 25 atm min^{-1} to 250 atm and finally 4.0 atm min^{-1} to 290 atm. The analysis time for all of these compounds on both the SB-cyanopropyl and the biphenyl phases was less than 25 min (Table 4.7). The authors concluded that both were suitable for the analysis of NSAIDs. This was demonstrated by the use of the SB-biphenyl column to analyse ibuprofen, ketoprofen and mefenamic acid in dosage forms (tablets or capsules).

In our own studies on the packed-column SFC of ibuprofen and naproxen (Roberts and Wilson, 1990) excellent results were obtained using an aminopropyl column (150 mm × 4.6 mm i.d., 5 μm aminopropyl Spherisorb) and CO_2–methanol (80:20, v/v) as mobile phase. The system was operated at 50 °C and 3000 psi. Typical chromatograms are illustrated in Figure 4.3. Unlike the situation encountered in HPLC, we found that the addition of an acidic modifier was not required in order to obtain retention and good peak shape. Furthermore, the amount of material that could be loaded on to an 'analytical' column without noticeable loss of peak shape was impressive (50–500 μg).

The high efficiency and good capacity of SFC indicated by these studies encouraged us to investigate the potential of SFC as a method for the

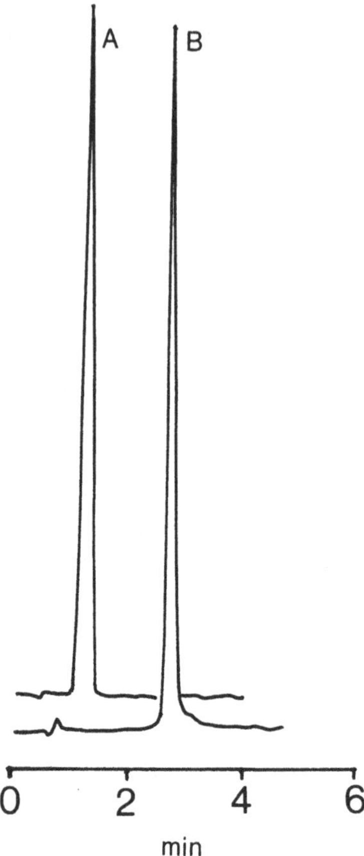

Figure 4.3 Packed-column SFC of (A) ibuprofen and (B) naproxen on aminopropyl-bonded silica gel (5 μm, 150 mm × 4.6 mm i.d.) using CO_2–methanol, 80:20 (v/v) at 3000 psi and 50 °C.

preparative isolation of drugs and metabolites from urine. Aspirin (acetylsalicylic acid) was used as a model compound. In man the major urinary metabolite of aspirin is the glycine conjugate of salicylic acid (*o*-hydroxyhippuric acid, salicyluric acid), although some salicylic acid and associated glucuronides together with various hydroxlated compounds (e.g. gentisic acid) have also been detected. A urine sample was obtained from a normal healthy volunteer following a single oral dose of 600 mg of the drug. Solid-phase extraction of aliquots of urine on to C_{18} bonded cartridges was used prior to SFC to concentrate and partially to purify the metabolites as well as to allow the sample to be dissolved in a suitable solvent for chromatography (methanol). SFC was performed on a diol-bonded silica (250 mm × 4.6 mm i.d., 5 μm Nucleosil) with CO_2–methanol (90:10) containing 0.1% trifluoroacetic acid to reduce peak tailing, at 3 ml

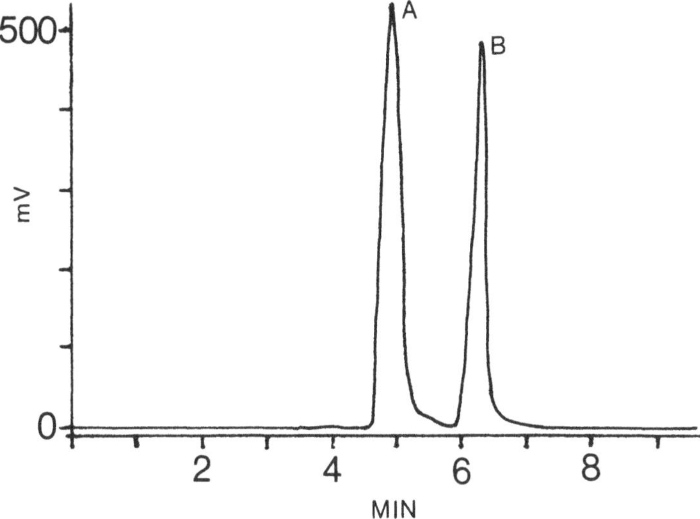

Figure 4.4 Packed-column SFC of (A) salicylic and (B) *o*-hydroxyhippuric acid on a diol-bonded silica gel (5 μm, 250 mm × 4.6 mm i.d.) with CO_2–methanol (containing 0.1% TEA) (90:10) at 3200 psi and 40 °C.

min^{-1} 40 °C and 3200 p.s.i. Under these conditions salicylic acid and *o*-hydroxyhippuric acid had retention times of 4.9 and 6.3 min, respectively (see Figure 4.4). The use of a cyano-bonded phase was also attempted, but without success.

Aliquots of the extracted urines were injected on to the column with CO_2 alone as mobile phase, and gradient elution was used to elute the analytes (0–20% methanol, with 0.1% TFA (v/v) over 5 min). Manual collection of the appropriate peak as it eluted from the detector (into methanol) facilitated the collection of sufficient material for identification by ^{1}H-NMR; the efficiency of this method was >95% with standards. A typical chromatogram for this type of work is shown in Figure 4.5.

The packed-column SFC of a range of polar benzoic acids including salicylic and salicylsalicylic acids has been studied by Berger and Deye (1991*a*) on cyanopropyl, diol and sulphonic acid phases (100 mm × 2 mm i.d., 5 μm (7 μm diol) Nucleosil). CO_2–methanol mixtures modified with either citric acid or trifluoroacetic acid were used as mobile phases. Typically the authors employed a temperature of 40 °C, and 200 bar.

4.2.13 Prostaglandins

SFE and linked packed-column SFC of the synthetic prostaglandin misoprostol, methyl (11a, 13*E*)-(±)-11,16-dihydroxy-16-methyl-9-oxoprost-13-en-1-oate, following SFE (see later) has been described (Roston *et al.*,

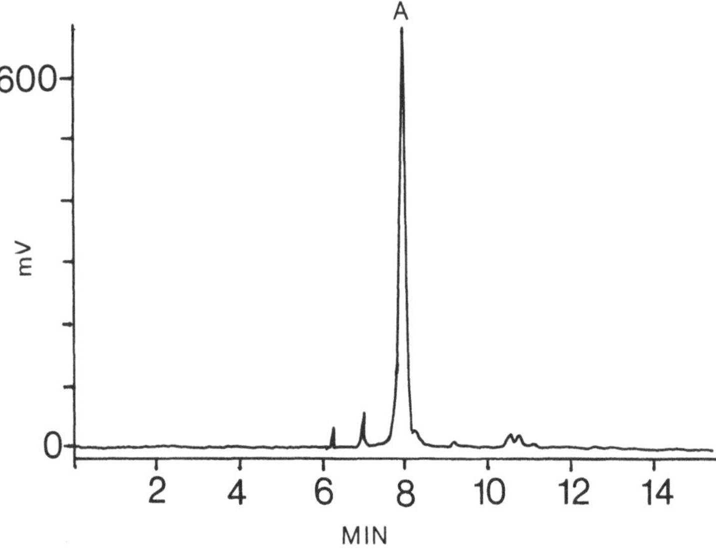

Figure 4.5 Packed-column SFC of a urine extract containing *o*-hydroxyhippuric acid (peak A). The sample was injected on to the column with CO_2 as the mobile phase. The analyte was then eluted using a gradient of methanol (containing 0.1% TFA) from 0 to 20% (v/v) over 5 min. The column temperature was maintained at 40 °C.

1992). SFC was accomplished using CO_2–methanol, 95:5, at a density of 0.8 g ml^{-1} and 70 °C on a cyano-bonded phase (200 mm × 1 mm i.d.) with UV detection at 195 nm. Under these conditions misoprostol eluted with a retention time of about 5 min.

Markides *et al.* (1986) demonstrated the use of capillary SFC for the separation of a mixture of six prostaglandins on a 12 m × 50 μm i.d. fused-silica column deactivated with cyanopropyl-methylhydrosilane and coated with a 50% cyanopropyl-methylpolysiloxane stationary phase. For this separation carbon dioxide was used as the mobile phase at 62 °C with density programming from 0.16 to 0.40 g ml^{-1} at 0.05 g ml^{-1} min^{-1} and thence from 0.40 to 0.87 g ml^{-1} at 0.007 g ml^{-1} min^{-1}. The different classes of prostaglandins were separated in the order A2, B2, E2 and F2α. The 5-*trans* isomer of F2 was separated from its parent under these conditions whilst methylated E2 epimers were partly resolved.

4.2.14 Purines and pyrimidines

SFC with UV detection (254 nm) was used by Mulcahey and Taylor (1990) to separate a number of purines and pyrimidines (permethrin, zidovudine, triluridine, mercaptopurine, triprolidine and pseudoephedrine). Separations were obtained by packed-column SFC on a cyano-bonded phase

(Deltabond CN, 250 mm × 4.6 mm i.d.) with CO_2–methanol as mobile phase. The SFC of permethrin, zidovudine, trifluidine and mercaptopurine at 60 °C was shown with CO_2 at 2ml min^{-1} initially modified with 150 μl of methanol (1.5 min) increasing to 450 μl min^{-1} over 2 min. Under these conditions peak tailing was noticeable for mercaptopurine. Both gradient elution and the addition of 0.001 M tetrabutylammonium bromide (TBA) (present in the methanol used as modifier) were used to obtain good separations. In particular, the TBA was needed to reduce excessive peak tailing for trimethoprim. An example was provided of the separation of permethrin, zidovudine and trimethoprim using gradient elution. The initial conditions were 2 ml min^{-1} CO_2 modified with 100 μl of methanol–TBA for 2 min and then increasing to 500 μl min^{-1} of methanol–TBA over 4 min at 60 °C. It was suggested that the improvements brought about by the inclusion of TBA were due to the modification of the stationary phase.

Triprolidine and pseudoephedrine were initially examined as their hydrochloride salts but were not eluted from the cyano column. However SFC of these compounds as their free bases was possible using similar conditions to those described above.

4.2.15 Sulphonamides

Sulphonamides have been the subject of a number of SFC studies. In an early study by Berry *et al.* (1986) a mixture of five sulphonamides (sulphamethoxypyridazine, sulphamerazine, sulphadiazine, sulphamethazine and sulphadimethoxime) were separated on a 100 mm × 4.6 mm i.d. column of 5 μm LiChrosorb. The mobile phase was CO_2 modified with 15% methanol at 4 cm^3 min^{-1}, 271 bar and 70 °C (analysis time was about 5 min). Detection was by UV (270 nm) and moving-belt MS. In a more detailed study from the same laboratory the packed-column SFC of ten sulphonamides, with UV (270 nm) and moving-belt or thermospray MS detection has been described (Perkins *et al.*, 1991*a*). The compounds studied were sulphadoxine, sulphamethazine, sulphamerazine, sulphadimethoxine, sulphadiazine, sulphaquinoxaline, sulphachloropyridazine, sulphathiazole, sulphamethoxypyridazine and sulphapyridine. Chromatography on two stationary phases was investigated, Spherisorb silica gel and Spherisorb aminopropyl-bonded silica gel (5 μm, 100 mm × 4.6 mm i.d. columns), with CO_2 modified with methanol, 2-methoxyethanol, acetonitrile, 2-propanol, dimethylformamide or propylene carbonate as the mobile phase. On the amino-bonded phase the baseline resolution of eight of the ten sulphonamides was possible; sulphadimethoxine and sulphamethoxypyridazine coeluted, and so did sulphamerazine and sulphapyridine. This result was obtained by using a gradient of methanol from 15% (4 min) to 25% (90 °C, 361 bar, 4 ml min^{-1}), with the increase in methanol concentration providing more rapid elution of the later-running peaks (sulphachloropyridazine and sulphathiazole) and better peak shape. The separation was achieved in 9.5 min.

The SFC of these compounds on a single 100 mm silica gel column proved unsatisfactory; however, linking two such columns in series provided good results and a different selectivity compared to the aminopropyl-bonded silica gel. Once again a gradient was employed with 12% methanol maintained for 5 min and then raised to 20% (v/v) (75 °C, 263 bar, 4 ml min^{-1}). Notwithstanding this difference in selectivity, however, the resolution of all ten sulphonamides was not achieved on silica either. Thus, sulphapyridine coeluted with sulphadiazine and sulphamerazine with sulphamethoxypyridine. It was noted that retention on the aminopropyl phase was much more sensitive to changes in the methanol content of the CO_2 than on the silica gel. Effects of the column pressure on the separations were also noted. Investigation of the use of the other organic modifiers showed that substitution with any of the solvents except dimethylformamide greatly increased the analysis time accompanied by peak broadening.

In another study Schmidt *et al.* (1988) used a total of five sulphonamides (sulphamethoxazole, sulphamethazine, sulphamerazine, sulphamethoxypyridazine and sulphamethizole) to investigate the effects of modifiers on packed and capillary SFC. In the case of capillary SFC best results were achieved using a 7 m × 50 μm i.d. SB-Cyanopropyl 25 (25% cyano, 25% propyl) coated fused-silica column at 100 °C with a mobile phase consisting of CO_2 containing 8% isopropanol (w/w). Detection was by UV at 254 nm. Nitrous oxide was also investigated as a mobile phase for the capillary SFC of these compounds. In the absence of an organic modifier no elution of the analytes was observed, whilst the addition of about 8% of isopropanol allowed elution but with poor peak shape. The results of these authors on packed columns (10 cm × 1 mm i.d. 5 μm Deltabond Cyano, and 15 cm × 1 mm i.d. 5 μm Spherisorb S5 ODS2) with either CO_2 or CO_2 premixed with 8% isopropanol were not particularly encouraging. The best results were obtained with the Spherisorb material with the isopropanol-modified mobile phase at 100 °C and injection at a pressure corresponding to a density of 0.34 g ml^{-1} for pure CO_2 and, after an isoconfertic period of 0.25 min, a programmed increase in density of 0.015 g ml^{-1} min^{-1}.

4.2.16 *Steroids*

Steroids have been the subject of a large number of SFC studies involving both capillary and packed columns, and with various types of detector (MS, infrared (IR), etc.). Capillary SFC of the steroid boldenone with

CE–MS detection was briefly reported by Lee *et al.* (1988), who demonstrated the superiority of this mode of ionisation over conventional EI–MS techniques. Chang and Lee (1988) have also published chromatograms showing the capillary SFC of androstanol isomers and oestrogens with CO_2 and FID.

The use of both capillary and packed-column SFC of five steroids with FID and IR detection was reported in two studies by Shah *et al.* (1988, 1989). The compounds involved (progesterone, testosterone, 17-hydroxyprogesterone, 11-deoxycortisol and cortisone) were separated on an SB-cyanopropyl-25 (10 m × 50 μm i.d., 0.25 μm film thickness) capillary column or 100 mm × 1.0 mm i.d. and 250 mm × 4.6 mm i.d. cyanopropyl-bonded phase silica packed columns (Deltabond and Spherisorb respectively) with CO_2 as the mobile phase. Separations were carried out at either 60 °C or 100 °C by pressure programming.

Later *et al.* (1986) reported the capillary SFC of such compounds as dexamethasone, betamethasone, cortisone, hydrocortisone and prednisolone. The conditions used for the separation of dexamethasone and betamethasone were as follows. SFC was performed on an 18 m × 50 μm i.d. SE-33 capillary column with CO_2 as the mobile phase. The temperature was 130 °C with pressure programming from 125 atm for 25 min then to 245 atm at 10 atm min^{-1}, and then to 275 atm at 5 atm min^{-1} and finally to 285 atm at 1 atm min^{-1}. Detection was by FID at 300 °C. The two compounds differ only by the α- or β-orientation of the C-16 methyl. Analysis time was about 30 min. As the authors point out, at the temperatures used for GC analysis epimerisation can occur leading to inaccurate quantification, but this does not happen under these SFC conditions. Cortisone and hydrocortisone were analysed under similar conditions with the pressure programme differing to the extent that, after the initial 25 min at 125 atm, the pressure was programmed up to 200 atm at 50 atm min^{-1}, then to 215 atm at 2 atm min^{-1} and finally to 245 atm at 5 atm min^{-1}. Analysis time was about 45 min. Prednisolone was identified in an ethyl acetate extract of horse urine following a ‘therapeutic’ dose of the compound. Analysis was performed on an 18 m × 50 μm i.d. methylpolysiloxane column with CO_2 as mobile phase at 130 °C and an initial pressure of 125 atm. After 10 min the pressure was increased to 245 atm at 50 atm min^{-1} and then to 320 atm at 10 atm min^{-1}. Analysis time appears to have been about 20 min, estimated from the published chromatogram. Once again, detection was by FID at 300 °C. Edlund and Henion (1989) have reported a similar study using packed column SFC–MS to identify prednisolone following intramuscular administration of 100 mg of the drug to the horse. Preparative TLC was used to partially purify the sample, which was then analysed on a 250 mm × 2 mm i.d. Spherisorb S3CN column. The mobile phase was CO_2-methanol, 98:2, at 3000 psi and 0.8 ml min^{-1} at a temperature of 70 °C. Under these conditions the compound eluted with a retention time of 4.8 min.

The analysis of steroids by capillary SFC following the formation of thiophosphinic ester derivatives was demonstrated by David and Novotny (1989) with detection using the phosphorus thermionic detector. A fused-silica column (10 m × 50 μm i.d.) coated with SE-33 (0.50 μm film thickness) was used. SFC-grade nitrous oxide was used as mobile phase. Thiophosphinic esters of a range of pregnanes and oestrogens were produced and detection limits in the low- or sub-nanogram range obtained. Pressure programming over the range 75–315 atm was used to elute the derivatives (8 atm min^{-1}) at a column temperature of 100 °C. Analysis times were of the order of 40–50 min per run. The methodology was successfully applied to both urine and plasma samples.

More recently Jagota and Stewart (1992*b*) have described the capillary SFC of methyltestosterone and five oestrogens (oestrone, equilin, α-oestradiol, β-oestradiol and δ-equilin) with FID detection. Chromatography on three different phases (5 m × 100 μm i.d. SB-methyl-100, 10 m × 50 μm i.d. SB-biphenyl-30 and 10 m × 50 μm i.d. SB-cyanopropyl-50) with CO_2 as mobile phase was investigated. Density programming, starting at an initial density of 0.700 g ml^{-1} (6 min), then by 0.0064 g ml^{-1} min^{-1} to 0.7500 g ml^{-1}, followed by 0.0100 g ml^{-1} min^{-1} to 0.8300 g ml^{-1} at 73 °C, with the SB-cyanopropyl-50 column gave the best results. Retention times under these conditions ranged from 12.9 min for oestrone to 19 min for δ-equilin. The methodology was applied to the analysis of the target compounds in three commercial dosage forms. The results obtained were not statistically different from those obtained by an HPLC procedure.

Packed-column SFC with UV detection (254 nm) of a mixture of testosterone, progesterone, nandrolone phenylpropionate, norgestrel, prednisolone, hydrocortisone, dexamethasone and ethynyloestradiol was briefly reported by Berry *et al.* (1986). The conditions employed were CO_2–methoxyethanol (80:20) at an initial flow rate of 2 cm^3 min^{-1} for 2.1 min, then 5 cm^3 min^{-1} at 195 bar and 75 °C. Methoxyethanol was used because of the inability of methanol to resolve all eight compounds when used as the modifier. The column used was silica (5 μm LiChrosorb, 100 mm × 4.6 mm i.d.) and the analysis time was about 6.5 min.

Recently, Berger and Deye (1991*b*) used steroids as probes to investigate the effects of column and mobile-phase polarity in packed-column SFC. These investigations employed cholest-4-en-3-one, progesterone, androstene-3,17-dione, testosterone, oestradiol, hydrocortisone and oestriol with UV detection at 220 nm. The columns used were 100 mm × 2 mm i.d. packed with C_{18} bonded, cyanopropyl, phenyl (5 μm, Hypersil phases), sulphonic acid (5 μm) and diol (7 μm, Nucleosil phases). The best results were obtained with polar stationary phases and methanol or isopropanol as organic modifier; acetonitrile and THF both gave poor results. Typical SFC conditions were 40 °C, 200 bar, CO_2 modified with 9.1% methanol at 1.1 ml min^{-1}.

A potentially interesting use of SFC to analyse steroids using light-scattering detection has been reported by Loran and Cromie (1990). However, apart from stating that chromatography was performed on a phenyl column (100 mm × 4.6 mm i.d.) few details of the mobile phases used, etc., were provided.

4.2.17 Tetracycline and oxytetracycline

The tetracycline antibiotics are high-molecular-weight polyfunctional compounds. Tetracycline, for example, has a molecular weight of 460 daltons and contains one amido, one diethylamine, two keto and six hydroxy functions. The capillary SFC of both oxytetracycline and tetracycline was reported (Later *et al.*, 1986) on a 13 m × 50 μm i.d. fused-silica column with SE-33 (0.65 μm film) as the stationary phase. The eluent was supercritical CO_2 at 100 °C with density programming from 0.25 g ml^{-1} (19 min isoconfertic) to 0.48 g ml^{-1} at 0.1 g ml^{-1} min^{-1} and then to 0.70 g ml^{-1} at 0.15 g ml^{-1} min^{-1}. Detection was by FID. Under these conditions both compounds coeluted. The packed and capillary column SFC of these analytes was also investigated by Schmidt *et al.* (1988). Packed-column SFC on either cyano or C_{18} bonded phases with either CO_2 or CO_2 containing 8% isopropanol proved to be unsuccessful. Use of a 3 m × 50 μm SB methyl-100 coated fused-silica capillary with carbon dioxide containing 8% isopropanol at 100 °C and density programming from 0.34 g ml^{-1} to 0.77 g ml^{-1} at a rate of 0.04 g ml^{-1} min^{-1} allowed the elution of oxytetracycline.

4.2.18 Xanthines

Edlund and Henion (1989) separated a number of xanthines, including caffeine, theophylline, pentoxyphylline (and an unknown metabolite isolated from horse urine by preparative thin layer chromatography) and diphylline (in that elution order) by packed-column SFC. The separation was accomplished by using a 100 mm × 4.6 mm i.d. Spherisorb S3CN column (3 μm) with CO_2–methanol, 95:5, at 1.5 ml min^{-1} as the mobile phase at a temperature of 70 °C. The inlet pressure was 4000 psi. Using EI SFC–MS, the unknown metabolite of pentoxyphilline was characterised as being *N*-demethylated. The availability of EI–MS in SFC as a means of obtaining mass spectral information was contrasted favourably with other approaches to SFC and HPLC–MS.

Berry *et al.* (1986) have also studied the packed-column SFC of the xanthines theophylline, theobromine and caffeine with both UV (270 nm) and MS detection. A good separation of all three compounds was achieved in about 4 min on silica (5 μm LiChrosorb, 100 mm × 4.6 mm i.d.) with a mobile phase consisting of CO_2–methanol, 88:12, at 2.5 cm^3 min^{-1}, 385 bar and 70 °C. Using a moving-belt interface, EI mass spectra were obtained from *m/z* 40 upwards, and gave excellent agreement with library spectra.

4.2.19 *Miscellaneous applications*

The capillary SFC of some polar ionophores used as veterinary antibiotics and a range of fat-soluble vitamins (including vitamins A, A acetate, K1, E, E acetate and K3, and provitamins D, D2 and D3) with CO_2 as mobile phase was reported by White *et al.* (1988). In addition to the SFC–MS of mitomycin C described earlier, Musser and Callery (1990) also demonstrated the microbore packed-column SFC–MS of three further anticancer drugs (diaziquone, thiotepa and cyclophosphamide) under the same conditions. Packed-column SFC for the veterinary antibiotics levamisole, furazolidine, chloramphenicol and lincomycin with MS detection was described by Perkins *et al.* (1991*b*). SFC was performed on amino-bonded silica gel with methanol-modified CO_2 as mobile phase. The same laboratory also described the use of SFE and packed-column SFC–MS–MS for the analysis of a further group of veterinary drugs (trimethoprim, hexestrol, diethylstilbestrol and dienestrol) in freeze-dried pig's kidney (Ramsey *et al.*, 1989).

In addition to their studies on the use of ion-pairing for the SFC of beta-blockers, Steuer *et al.* (1990) also separated the drug spirapril and some degradation products on a diol column (100 mm × 4.6 mm i.d.) with CO_2–methanol (80:20) containing 20 mM TBA and 20 mM acetic acid. These authors also described the resolution of the drug isradipin and ten degradation products. SFC was carried out using a silica column (Spheri 5, 5 μm, 400 mm × 4.6 mm i.d.) with CO_2–methanol (96:4, v/v), at 50 °C and 128 bar, as eluent. In both instances the benefits of pressure programming as a means of improving separations were illustrated.

SFC was used in combination with SFE for the analysis of the steroid triamcinolone acetonide in dermatological patches (Edwardson and Gardner, 1990). Chromatography was apparently undertaken by using a 200 mm × 4.6 mm i.d. column of LiChrosorb silica (10 μm), but no further details were provided.

Di Maso *et al.* (1990) have briefly described the use of SFC for the analysis of sorbitan trioleate in metered dose inhalers, where it is used to keep drugs dispersed in the propellant and to lubricate the actuator valve. SFC was performed on a C_{18} bonded packed column (5 μm Rexchrom, 100 mm × 2.1 mm i.d.) with CO_2 as the mobile phase (2400 psi at 40 °C) with detection by FID. The methodology developed was used to analyse material from a number of formulations.

4.3 Drugs of abuse

Whilst not always pharmaceuticals as such, drugs of abuse are clearly often related to them. Without going into the literature in detail it is perhaps worth noting that such compounds have been the subject of studies in SFC. The paper by Later *et al.* (1986) describes the capillary SFC of tetrahydrocannabinol and six metabolites, as well as other compounds. Crowther and Henion (1985) demonstrated the use of packed-column SFC–MS for cocaine, codeine, methocarbamol, phenylbutazone and oxyphenylbutazone. Chromatography was performed on a range of column packings including silica, amino, cyano and diol phases packed in 200 mm × 2.1 mm i.d. columns with methanol as a modifier. These authors concluded that SFC offered advantages over HPLC in terms of speed of analysis and column re-equilibration. The use of SFC with UV (packed column), FID and MS (capillary column) detection for the separation and identification of a cocktail of substances including controlled drugs and substances used as diluents was described by Mackay and Reed (1991). The compounds analysed were phenacetin, phenobarbitone, methaqualone, *N*-phenyl-2-naphthylamine, acetylcodeine, diamorphine, papaverine, narcotine and phenolphthalein.

4.4 Supercritical fluid extraction

Compared to SFC, where there are a good number of literature examples of applications to drugs, the situation with regard to supercritical fluid extraction (SFE) is that to date comparatively few examples are available. This is no doubt a reflection, at least in part, of the lack of suitable commercially available SFC apparatus rather than an indication that SFE is unsuited to the extraction of pharmaceuticals. Some examples of the uses of SFE will now be provided.

Certainly there appears to be potential for SFE in the isolation and manufacture of drugs as an alternative to milling (Larson and King, 1986) and to the extraction of fermentation products. In particular SFE was shown to be suitable for the extraction of mevinolin, a metabolite of the fungus *Aspergillus terreus* from a freeze-dried fermentation broth.

The potential of SFE for extracting drugs from plant materials has also been demonstrated for papaverine from *Papaver somniferrum* and quinine from the West African herbs 'lemon grass' and 'dogonyaro leaves' (Ndiomu and Simpson, 1988).

SFE has also been used for the analysis of drugs in pharmaceutical dosage forms. Attempts to use SFE for the extraction of triamcinolone from dermatological patches using CO_2 modified with 0, 5 or 10% methanol (80 °C, 5 min) proved largely unsuccessful (Edwardson and Gardner,

1990). Recoveries of only 10% were achieved. The SFE of misoprostol, a synthetic prostoglandin, from a hydroxypropyl methylcellulose dispersion has been described with extraction efficiencies of about 65% (Roston, 1992).

In so far as biomedical applications of SFE for drug analysis are concerned, the technique has been applied to the extraction of morphine from freeze-dried serum samples (Ndiomu and Simpson, 1988) spiked with morphine (200 $\mu g\ ml^{-1}$). Recoveries of 96.7% (RSD 3.2%) were obtained, compared with 92.2% for solid-phase extraction (SPE). However, the SPE extraction could be performed without the need for freeze-drying and was therefore much less time-consuming. In the same study SFE was also used successfully for the analysis of morphine in placental samples.

A further interesting example of the potential of SFE in drug analysis in biological samples is provided by the work of Liu and Wehmeyer (1992), who combined solid-phase extraction with recovery of the extracted analyte by SFE; in this case flavone in dog plasma was used as a model. Plasma samples were spiked with the analyte at 10–250 ng ml^{-1} and then extracted on to a C_{18} bonded SPE phase. When the analyte had been extracted, the C_{18} cartridge was placed in an SFE extraction cell and eluted with CO_2–methanol (95:5) at 50, 100, 150 and 200 bar. At 150 bar, recoveries ranging from 100.8% for 10 ng ml^{-1} to 91.3% at 250 ng ml^{-1} were obtained. It was apparent, however, that these conditions also eluted unwanted material from the cartridges themselves. This effect could be eliminated by removing the packing material from the cartridge prior to SFE, but this procedure would clearly be time-consuming if carried out routinely. The accuracy of SPE–SFE was considered to be comparable to that of conventional SPE methods.

4.5 Conclusions

As can be seen from the examples provided, SFC seems to be broadly applicable to a wide range of drugs and pharmaceuticals ranging from non-polar steroids and prostaglandins to polar cephalosporins and acidic NSAIDs. Indeed, as illustrated above there exists a good number of examples of the use of both packed-column and capillary SFC in the literature. As a generalisation, it appears to be the case at the moment that SFC of pharmaceuticals and drugs is almost universally performed using CO_2 with methanol as the organic modifier of choice.

Clearly, SFC has some promise; however, even a cursory examination of the literature shows that many of the examples not only are of recent origin but also demonstrate the potential applications of SFC rather than fully developed methods. In this area, therefore, SFC remains in its infancy.

In contrast to SFC, SFE does not yet appear to have been exploited to any great extent. This probably does not indicate that SFE is inapplicable to drugs and pharmaceuticals but merely reflects the lag time between the availability of the appropriate apparatus and the preparation of texts for publication.

Even though SFC and SFE can clearly be applied in these areas it is still difficult to forecast what the future holds for them. This is because, as indicated in the introduction, the widespread adoption of these techniques depends to no small degree upon the availability of competitively priced and reliable equipment. Given the excellent results that can be obtained by SFC, it is to be hoped that developments in instrumentation will continue so as to ensure the more widespread deployment of this methodology.

References

Berger, T.A. and Deye, J.F., (1991*a*), Separation of hydroxybenzoic acids by packed column supercritical fluid chromatography using modified fluids with very polar additives. *J. Chromatogr. Sci.*, **29**, 26–30.

Berger, T.A. and Deye, J.F. (1991*b*), Effects of column and mobile phase polarity using steroids as probes in packed column supercritical fluid chromatography. *J. Chromatogr. Sci.*, **29**, 280–285.

Berry, A.J., Games, D.E. and Perkins, J.R., (1986), Supercritical fluid chromatographic and supercritical fluid–mass spectrometric studies of some polar compounds. *J. Chromatogr.*, **363**, 147–158.

Chang, H.-C.K. and Lee, M., (1988), Atlas of Chromatograms, SFC38 and SFC39. *J. Chromatogr. Sci.*, **26**, 298.

Crowther, J.B. and Henion, J.D., (1985), Supercritical fluid chromatography of polar drugs using small-particle packed columns with mass spectrometric detection. *Anal. Chem.*, **57**, 2711–2716.

David, P.A. and Novotny, M., (1989), Analysis of steroids by capillary supercritical fluid chromatography with phosphorous-selective detection. *J. Chromatogr.*, **461**, 111–120.

Di Maso, M., Purdy, W.C., McClintock, S.A. and Cotton, M.L., (1990), Determination of sorbitan trioleate in metered-dose inhalers by supercritical-fluid chromatography. *J. Pharm. Biomed. Anal.*, **8**, 303–305.

Edlund, O.P. and Henion, J.D., (1989), Packed-column supercritical fluid chromatography/mass spectrometry via a two-stage momentum separator. *J. Chromatogr. Sci.*, **27**, 274–282.

Edwardson, P.A.D. and Gardner, R.S., (1990), Problems associated with the extraction and analysis of triaminoline acetonide in dermatological patches. *J. Pharm. Biomed. Anal.*, **8**, 935–938.

Jagota, N.K. and Stewart, J.T. (1992*a*), Separation of non-steroidal anti-inflammatory agents using supercritical fluid chromatography. *J. Chromatogr.*, **604**, 255–260.

Jagota, N.K. and Stewart, J.T., (1992*b*), Supercritical fluid chromatography of selected oestrogens. *J. Pharm. Biomed. Anal.*, **10**, 667–674.

Janicot, J.L., Caude, M. and Rosset, R., (1988), Separation of opium alkaloids by carbon dioxide sub- and supercritical fluid chromatography with packed columns: application to the quantitative analysis of poppy straw extracts. *J. Chromatogr.*, **437**, 351–364.

Lane, S.J., (1988), SFC-MS in the pharmaceutical industry, In *Supercritical Fluid Chromatography*, Smith, R.M. (Ed.), RSC Chromatography Monographs, Royal Society of Chemistry, London, pp. 175–202.

Larson, K.A. and King, M.L., (1986), Evaluation of supercritical fluid extraction in the pharmaceutical industry. *Biotech. Prog.*, **2**, 73–82.

Later, D.W., Richter, B.E., Knowles, D.E. and Anderson, M.R., (1986), Analysis of various classes of drugs by supercritical fluid chromatography. *J. Chromatogr. Sci.*, **24**, 249–253.

Lee, C.R., Porziemsky, J.-P., Aubert, M.-C. and Krstulovic, A.M., (1991), Liquid and high-pressure carbon dioxide chromatography of β-blockers: Resolution of the enantiomers of nadolol. *J. Chromatogr.*, **539**, 55–69.

Lee, E.D., Hsu, S-H. and Henion, J.D., (1988), Electron-ionisation-like mass spectra by capillary supercritical fluid chromatography/charge exchange mass spectrometry. *Anal. Chem.*, **60**, 1990–1994.

Loran, J.S. and Cromie, K.D., (1990), An evaluation of the use of supercritical fluid chromatography with light scattering detection for the analysis of steroids. *J. Pharm. Biomed. Anal.*, **8**, 607–611.

Liu, H. and Wehmyer, K.R., (1992), Solid phase extraction with supercritical fluid elution as a sample preparation technique for the ultratrace analysis of flavone in blood plasma. *J. Chromatogr.*, **577**, 61–67.

Mackay, G.A. and Reed, G.D., (1991), The application of capillary SFC, packed column SFC and capillary SFC–MS in the analysis of controlled drugs. *J. High Resolut. Chromatogr.*, **14**, 537–541.

Markides, K.E., Fields, S.M. and Lee, M.L., (1986), Capillary supercritical fluid chromatography of labile carboxylic acids. *J. Chromatogr. Sci.*, **24**, 254–257.

Mount, D.L., Patchen, L.C. and Churchill, F.C., (1990), Determination of mefloquine in blood by supercritical fluid chromatography with electron capture detection. *J. Chromatogr.*, **527**, 51–58.

Mulcahey, L.J. and Taylor, L.T., (1990), Gradient separation of pharmaceuticals employing supercritical CO_2 and modifiers. *J. High Resolut. Chromatogr.*, **13**, 393–396.

Musser, S.M. and Callery, P.S., (1990), Supercritical fluid chromatography/chemical ionisation/mass spectrometry of some anticancer drugs in a thermospray ion source. *Biomed. Environ. Mass Spec.*, **19**, 348–352.

Ndiomu, D.P. and Simpson, C.F., (1988), Some applications of supercritical fluid extraction. *Anal. Chim. Acta.*, **213**, 237–243.

Niessen, W.M.A., Bergers, P.J.M., Tjaden, U.R. and Van Der Greef, J., (1988), Phase-system switching as an on-line sample pretreatment in the bioanalysis of mitomycin C using supercritical fluid chromatography. *J. Chromatogr.*, **454**, 243–251.

Niessen, W.M.A., Tjaden, U.R. and Van Der Greef, J., (1989), Bioanalytical applications of supercritical fluid chromatography. *J. Chromatogr.*, **492**, 167–188.

Perkins, J.R., Games, D.E., Startin, J.R. and Gilbert, J., (1991*a*), Analysis of sulphonamides using supercritical fluid chromatography and supercritical fluid chromatography–mass spectrometry. *J. Chromatogr.*, **540**, 239–256.

Perkins, J.R., Games, D.E., Startin, J.R. and Gilbert, J., (1991*b*), Analysis of veterinary drugs using supercritical fluid chromatography and supercritical fluid chromatography-mass spectrometry. *J. Chromatogr.*, **540**, 257–270.

Ramsey, E.D., Perkins, J.R., Games, D.E. and Startin, J.R., (1989), Analysis of drug residues in tissue by combined supercritical-fluid extraction–supercritical fluid chromatography–mass spectrometry–mass spectrometry. *J. Chromatogr.*, **464**, 353–364.

Roberts, D.W. and Wilson, I.D., (1990), Bioanalytical supercritical fluid chromatography, In *Analysis for Drugs and Metabolites*, Reid, E. and Wilson, I.D. (Eds), Royal Society of Chemistry, Cambridge, pp. 257–263.

Roston, D.A., (1992), Supercritical fluid extraction–supercritical fluid chromatography for analysis of a prostaglandin:hpmc dispersion. *Drug Dev. and Ind. Pharmacy.*, **18**, 245–255.

Ruane, R.J., Tomkinson, G.P. and Wilson, I.D., (1990), The detection of [^{14}C]-propranolol following supercritical fluid chromatography using in-line radioactivity detection. *J. Pharm. Biomed. Anal.*, **8**, 1091–1093.

Schmidt, S., Blomberg, L.G. and Campbell, E.R., (1988), Modifier effects on packed and capillary columns in supercritical fluid chromatography. *Chromatographia*, **25**, 775–780.

Shah, S., Ashraf-Korassani, M. and Taylor, L.T., (1988), Demonstration of an optimised flow cell SFC–FTIR interface. *Chromatographia*, **25**, 631–635.

Shah, S., Ashraf-Korassani, M. and Taylor, L.T., (1989), Normal-phase liquid and supercritical-fluid chromatographic separations of steroids with Fourier-transform infrared spectrometric detection. *Chromatographia*, **26**, 441–448.

Siret, L., Bargmann, N., Tambute, A. and Caude, M., (1992), Direct enantiomeric separation of β-blockers on ChyRoSine-A by supercritical fluid chromatography: Supercritical carbon dioxide as a transient *in situ* derivatising agent. *Chirality*, **4**, 252–262.

Smith, R.M. and Sanagi, M.M., (1988), Application of packed column supercritical fluid chromatography to the analysis of barbiturates. *J. Pharm. Biomed. Anal.*, **6**, 837–841.
Smith, R.M. and Sanagi, M.M., (1989*a*), Supercritical fluid chromatography of barbiturates. *J. Chromatogr.*, **481**, 63–69.
Smith, R.M. and Sanagi, M.M., (1989*b*), Packed-column supercritical fluid chromatography of benzodiazepines. *J. Chromatogr.*, **483**, 51–61.
Steuer, W., Schindler, M., Schill, G. and Erni, F., (1988), Supercritical fluid chromatography with ion-pairing modifiers: Separation of enantiomeric 1,2-aminoalcohols as diastereomeric ion pairs. *J. Chromatogr.*, **447**, 287–296.
Steuer, W., Baumann, J. and Erni, F., (1990), Separation of ionic drug substances by supercritical fluid chromatography. *J. Chromatogr.*, **500**, 469–479.
Veuthey, J.-L. and Haerdi, W., (1990), Separation of amphetamines by supercritical fluid chromatogrpahy. *J. Chromatogr.*, **515**, 385–390.
White, C.M., Gere, D.R., Boyer, D., Pacholec, F. and Wong, L.K., (1988), Analysis of pharmaceuticals and other solutes of biochemical importance by supercritical fluid chromatography. *J. High Resolut. Chromatogr.*, **11**, 94–98.
Wong, S.H.Y. and Dellafera, S.S., (1990), Supercritical fluid chromatography for therapeutic monitoring and toxicology; methodological considerations for open capillary tubular column for the analysis of phenobarbital in serum. *J. Liq. Chromatogr.*, **13**, 1105–1124.

5 Applications of supercritical fluids in polymer analysis

D.E. KNOWLES and T.K. HOGE

5.1 Introduction

The use of supercritical fluids for chromatography and/or extraction is applicable to a wide variety of compound classes. Of all the application areas that have been investigated to date, polymers have probably benefitted from the use of supercritical fluids more than any other. Polymers and polymer additives that are found in foods, textiles, surfactants, cosmetics, the environment, and the petroleum industries all have become areas where supercritical fluid chromatography (SFC) and supercritical fluid extraction (SFE) are presently finding widespread acceptance and use.

The power of supercritical fluid technology rests in density programming. While supercritical fluids have liquid-like densities, their solvent strength, in some cases, is orders of magnitude greater than the solvent strength of liquids. Because of the solvent strength, small changes in pressure can often result in large changes in solubility (Sie and Rijnders, 1967). As will be shown, manipulating the solubility will allow for separation of oligomers, isomers, secondary and often tertiary distributions of the polymer being analyzed. Also, since separations are done mainly utilizing solubility, samples do not have to be volatile to be analyzed. To better understand the importance of this, a comparison of present analytical techniques for polymers is important.

Gas chromatography (GC) is often the first technique of choice. Gas chromatography works well for those polymers that are of low molecular weight (typically under 1000 daltons) and volatile, and dissolve in low polarity solvents. Method development is straightforward and fast. Temperature programming is used to separate out the components of the polymer. Optimization of the chromatography in GC is usually only a matter of trying a couple of different temperature ramps or selecting a different stationary phase. Because of the high diffusivity in gases, analysis times are very short. High performance liquid chromatography (HPLC) employs a solvating mobile phase, high densities, and high viscosities. Polymers that are non-volatile, of high molecular weight, and/or polar are more often separated using HPLC as the analysis technique. Method development using HPLC is often time-consuming. Parameters that must be considered in HPLC include column type, mobile phase composition, and the gradient to be used. Run times are typically longer using HPLC than with GC or SFC. Preferably, the sample should be chromophoric in nature, since the use of a universal detector is still not optimized for HPLC. Size exclusion chromatography (SEC) and gel permeation chromatography (GPC) are techniques used to determine molecular-weight distributions, but not oligomeric separations. Ion chromatography (IC) is also a technique that is used for polymer samples such as ionic surfactants and some additives.

Supercritical fluids are those fluids that meet important criteria. They must be above their critical temperature (T_c), above their critical pressure (P_c), and also they must be strong enough to cause solute migration, or possess solvating strength. Supercritical fluids have diffusivities that are more GC-like, viscosities lower than liquids and densities that are more liquid-like. The diffusion and viscosities of a supercritical fluid allow for fast, efficient GC-like separations with solvating properties similar to HPLC. In a simplistic sense, SFC can be thought of as doing GC with a solvating mobile phase (thereby expanding GC capabilities) or expanding HPLC analysis capabilities by being able to use universal detectors. Method development in SFC, while often slower than in GC, is faster than in HPLC. Pressure or density programming is typically employed. As one increases the pressure (under isothermal conditions) the density increases, and this increase in density also increases the solvent strength (Richter, 1989). This is analogous to temperature programming in GC. The method-development 'mind-set' is similar. In GC, if one wants to maximize separations the pressure ramp is slowed down; in SFC to maximize separations the pressure ramp is slowed. The difference is that in GC this is a volatility effect, while in SFC the change is in solubility.

For extraction or chromatography, the choice of the mobile phase is of paramount importance. The vast majority of work that has been done on polymers utilizes carbon dioxide as the mobile phase. CO_2 has a low critical temperature (31 °C), a low critical pressure (72.9 atm), and high solvating strength. It is non-toxic, inexpensive, colorless, non-flammable, and does not respond in a flame ionization detector (FID). Thus, with CO_2, universal detection is available. It works well in a UV detector, since its UV cutoff (190 nm) is lower than those of many of the solvents used in HPLC. CO_2 is usually the solvent of choice for those polymers that have molecular weights below 10 000 daltons, and are fairly non-polar and non-ionic. If the polymer being analyzed contains high polarity, primary amines, sugars, more than six hydroxy groups, or ionic species, it will not be soluble in CO_2. For those polymers that are not soluble in CO_2, there are two possibilities: (1) use a different pure mobile phase; or (2) add an organic modifier to the CO_2 to change the solubility of the fluid or modify the stationary phase in SFC, or to swell the matrix when doing SFE. The decision on which avenue to take depends on the analyte and the supercritical fluid technique being used. There are three supercritical fluid techniques that are presently being used. These are supercritical fluid chromatography, off-line supercritical fluid extraction, and on-line supercritical fluid extraction coupled to a GC, HPLC, or SFC.

5.2 Supercritical fluid chromatography

The first growth of supercritical fluids as an aid to the analysis of polymers was through its use in chromatography. Using the supercritical fluid as a mobile phase in SFC gives rise to many possibilities that previously did not exist. Because of the solvating nature of the mobile phase, polymers that are thermally labile or are non-volatile can readily be analyzed under supercritical conditions.

Another important factor in SFC is the availability of a wide range of detectors. Most of the commonly used GC and HPLC detectors have been interfaced to SFC. The most common detector used in SFC is the FID. Because it is a universal detector, the analytes of interest do not need to have chromophores as they do for many HPLC analyses. The UV detector that is commonly used in HPLC also works well with supercritical fluids. The flow cell must be adapted to allow operation under the high pressures used in SFC. Mass spectrometers (MS), Fourier transform infrared (FT-IR), nitrogen phosphorus detectors (NPDs), electron capture detectors (ECDs), and sulfur chemiluminescence detectors (SCDs) are forms of detection often used in SFC.

5.2.1 Method development

To begin polymer analysis by SFC, decisions have to be made concerning the mobile phase, detectors, injection scheme, and stationary phase that are to be used.

5.2.2 Mobile phase

Choosing the mobile phase to use is dependent on first the polymer that is to be analyzed and second the detector that is being employed. Carbon dioxide is, by far, the most often used mobile phase for polymer analysis. Besides its low critical conditions, CO_2 does not respond in a flame and therefore the FID can be used. CO_2 can solvate polymers of up to 10 000 daltons, or slightly higher if fluorinated. It is possible to solvate higher-molecular-weight polymers with other mobile phases. For example, Fujimoto *et al.* (1989) have reported the elution of polystyrenes that have molecular weights of almost 3 000 000 daltons. To get the polystyrene into solution it was necessary to use methylene chloride as the supercritical fluid. For those analytes where CO_2 is not the fluid of choice, often an organic modifier is added to the CO_2. Modifying CO_2 with an organic modifier does several things. The solvent strength is changed, the polarity can also change depending on the modifier used, and the critical constants are affected. In SFC often less than 10% modifier by weight is added to the CO_2. In practice, this has little effect on the solvent strength. Since CO_2 is a relatively non-polar solvent, increasing the polarity may be necessary for some of the more polar polymers. Again, this can be done by simply adding a small amount of a polar organic modifier. Whenever modifiers are added, either to increase solvent strength or to change the polarity of the fluid, the critical constants also change. For example, methanol (MeOH) is a popular modifier for CO_2. Methanol has a critical temperature of 240.5 °C and a critical pressure of 78.9 atm. With the small difference between CO_2 and methanol in critical pressure, there will not be a large effect on the critical pressure of the mixture. However, with the extremely high critical temperature for methanol, the critical temperature of the mixture will be much higher than 31 °C. In most cases, it may not be detrimental to operate at this higher temperature. However, if the analyte being chromatographed is thermally labile or can polymerize at low temperatures (e.g. isocyanates), then this becomes a concern. In capillary columns, the vast majority of work published has been done using a pure mobile phase. A large percentage of the packed-column SFC work found in the literature has used modified mobile phases (Lee and Markides, 1990). Capillary columns are extremely well deactivated, while packed columns have many residual silanol groups in the packing material. These active sites in a packed column often have detrimental effects on the elution of analytes. To lessen the problems caused by active sites, modified CO_2 can be used as the mobile phase. The organic modifier will mask the silanol groups, thereby allowing for analysis of those analytes that may previously have interacted with the stationary phase. Whenever the mobile phase has had an organic modifier added to it, an FID can not be used. With any modifier (with the exception of low concentrations of formic acid, or freons) an unstable baseline or baseline rise will occur with the FID. Therefore, ultraviolet (UV), fluorescence or MS would become the detector of choice.

5.2.3 Detectors

The FID is the most commonly used detector in SFC. Use of this detector allows SFC to have a universal detector. The UV detector is often the detector of choice when modified fluids are used, or in cases where it is advantageous to be able to detect selectively an analyte that has chromophores from a series of oligomers that may not. For analyte identification purposes the MS can be used. SFC–MS is also commercially available and is much more routinely used than is HPLC–MS. While other detection schemes have been used for polymer analysis, these three detectors represent the detection schemes for the vast majority of the methods that have been published to date (Later *et al.*, 1988*a*).

5.2.4 *Injection*

There are four injection techniques that are routinely used in SFC for polymer analysis. These are split injection, timed-split injection, direct injection, and on-line extraction. In a split injection, the entire sample loop contents are injected. An external split restrictor is then used to eliminate most of the contents of the sample loop so that only a small portion of the contents of the sample loop goes on to the column. This type of injection is typical of that used in GC. Timed-split injection was first used in HPLC by Harvey and Stearns (1984). In this technique, the entire sample loop is filled. By controlling the timing of the injector, only a small portion of the contents of this loop is injected and goes on to the columnn. A direct injection is similar to a timed-split injection and is routinely used in packed-column SFC. In a direct injection, the entire contents of the sample loop are injected and go on to the column (Richter *et al.*, 1988). On-line extraction is a very powerful technique and is used to eliminate solvent or matrix effects. When cryogenic trapping of the extracted material is used the analytes are focused or concentrated and the sensitivity can be drastically improved (Richter *et al.*, 1986). For polymers, and especially higher-molecular-weight polymers, the timed-split and direct injections are the most often used.

5.2.5 *Stationary phase*

For SFC, work has been done using capillary, packed, and packed capillary columns. The capillary column is the column of choice for high resolution and high chromatographic efficiency, and when higher-molecular-weight polymers are being analyzed. This allows for the maximum utilization of density and solubility. As the pressure (density) increases, the solubility increases and higher-molecular-weight polymers become soluble. In capillary columns there is only a 1–3% pressure drop over the column. Using a capillary column, Campbell (1991) has recently shown oligomeric separation of polybutenes up to C_{146} with CO_2 as the mobile phase. Packed columns allow for more material to be injected. They also offer more plates per unit time than do capillary columns. The most often used packed columns in polymer analysis are the microbore HPLC columns. Typically, there is a 20–50% pressure drop over these columns. Because of this, with a packed column the range of pressure able to be used may not be as great as with capillary, minimizing the solvation of high-molecular-weight polymers. C_{108} is the highest carbon number that has been separated on packed columns with CO_2 mobile phase. Because of the residual silanol groups, packed columns are inherently more polar than capillary columns. Being more polar, packed columns are often the columns used for the more polar polymers. Packed capillaries offer more loadability than capillaries and often less pressure drop than packed columns. They are also hard to deactivate and presently are not readily used in SFC.

As has been stated previously, the power of SFC lies in increased solvent strength of the mobile phase. The stationary-phase selectivity of columns is also a strong factor for optimizing separations. Figure 5.1 shows the separation power that can be enhanced by changing the selectivity of the column. The two chromatograms utilized the same conditions, sample, concentration, column length, column internal diameter, and film thickness of the stationary phase. The only difference was the stationary phase being used. A methyl column is a non-polar stationary phase, while the biphenyl column is slightly more polar. On the methyl column a homologous series is seen. This same series is apparent on the biphenyl column, but because of the polarizability of the stationary phase, other components

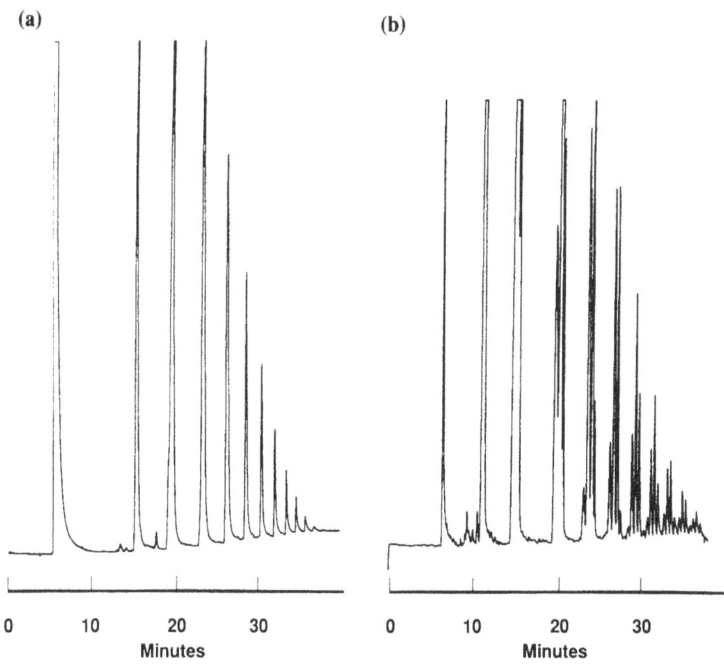

Figure 5.1 Industrial polymers using two different stationary phases: (a) SB-Methyl-100; 10 m × 50 μm i.d.; (b) SB-Biphenyl-30; 10 m × 50 μm i.d. Chromatography conditions: mobile phase, CO_2; FID at 350 °C; split injection with 20:1 split ratio; density program: 0.25 g ml^{-1} held for 10 min, then ramped at 0.015 g ml^{-1} min^{-1} to 0.76 g ml^{-1}.

(isomers, secondary distributions) within the oligomers can be baseline-resolved. Optimizing the chromatography is often accomplished by changing the stationary phase.

5.3 Applications of SFC for polymers

The success and growth of any analysis technique depends on the applications that are possible with the technology. In this section, examples of supercritical fluid chromatograms for the analysis of polymers and polymer additives will be shown.

5.3.1 α-Olefins

α-Olefins are used in a wide variety of products including foods, shampoos, soaps, detergents, inks, and lubricants, and as precursors to other polymers. The carbon number of the olefin is used to determine what product the olefin is used for. For example, shampoos typically contain C_{16} and C_{18}-α-olefins. Often the list of ingredients on the shampoo bottle will state that there are C_{16},C_{18}-α-olefins and residuals present. Typically an olefin stream is fractionated by hydrocarbon number. The mid-range olefins (C_{22}–C_{30}) have been difficult to analyze by chromatographic means. They are temperature-sensitive and non-chromophoric. However, with SFC these olefins are routinely analyzed. Figure 5.2 shows a chromatogram of C_{24}–C_{48}-α-olefins. The ability to use low temperatures and an FID allows for the analysis of these olefins. The major components in the chromatogram are the α-olefins. The secondary distribution is actually the β-position. At around C_{48}, the distribution shifts to the β-position from the α-position.

5.3.2 Polyglycols

Extensive work has been published on the analysis of polyglycols. Polyglycols are widely used as starting materials for other polymers such as polyurethanes. Other popular uses for polyglycols include such diverse applications as lubricants, surfactants, cosmetics, food products and pharmaceuticals. Gas chromatography can be used to analyze some of the low-molecular-weight glycols. High temperature gas chromatography (HTGC) can increase this molecular-weight range, but even at 400 °C there is a limit to the molecular weight that can be eluted. Many of these polymers do not inherently contain UV-absorbing moieties, so HPLC is often insufficient for glycols. Richter (1989) has shown elution of polyglycols of over 5000 daltons with SFC. Chester *et al.* (1990) have shown that higher pressures can be used to separate poly(ethylene glycol) 2000 with oligomeric separa-

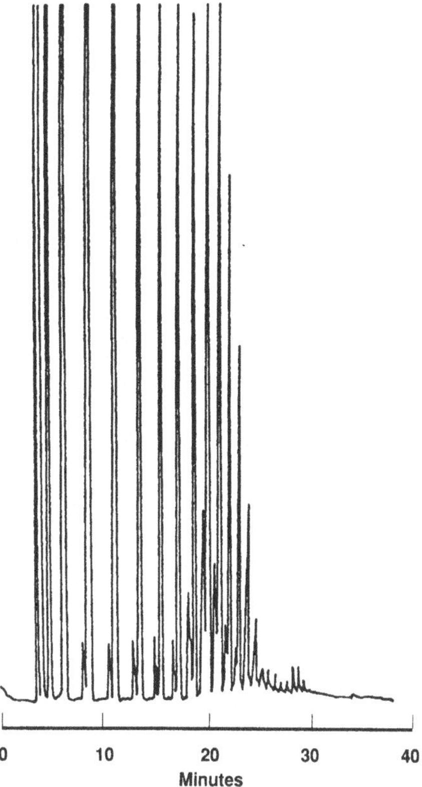

Figure 5.2 Supercritical fluid chromatogram of C_{22}–C_{48} α-olefins. Conditions: CO_2 at 100 °C; FID at 350 °C; timed-split injection at 0.050 s; concentration: 50 mg ml^{-1}; column: 4.5 m × 50 μm i.d. SB-Methyl-100; density program: 0.25 g ml^{-1} held for 5 min, then ramped to 0.76 g ml^{-1} at 0.015 g ml^{-1} min^{-1}.

tions. Knowles *et al.* (1988) have demonstrated the ability to analyze lot-to-lot comparisons of polyglycols using SFC as a quality control (QC) technique for these polymers. For example, using SFC on different lots of polyglycols will ensure that the lot-to-lot glycol distribution remains the same. With this information, the polyglycols can confidently be used as starting materials, knowing that the final product performance will not fluctuate due to a change between different lots of material being used.

5.3.3 Ethoxylated alcohols

Ethoxylated alcohols are made by the addition of ethylene oxide (EO) to aliphatic alcohols. As non-ionic surfactants, they are used in detergents, as emulsifiers, and as foaming and wetting agents. To analyze these by HPLC requires the surfactant to be derivatized because they do not contain UV-absorbers. Elution by GC is limited to low levels of ethoxylation. HTGC

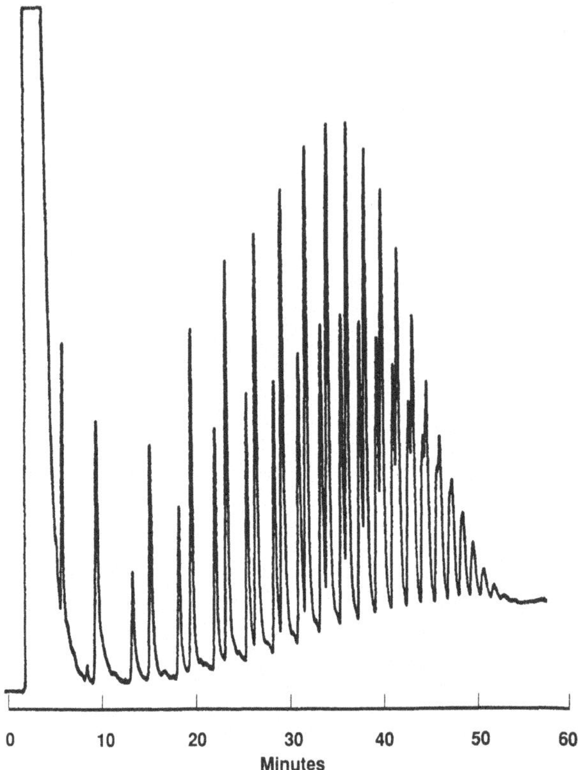

Figure 5.3 Supercritical fluid chromatogram of C_{12},C_{14} ethoxylated alcohols. Conditions: mobile phase, CO_2 at 100 °C; FID at 400 °C; column: 5 m × 50 μm i.d. SB-Biphenyl-30; timed-split injection at 0.100 s; concentration: 100 mg ml^{-1}; pressure program: 125 atm held for 10 min, then ramped at 300 atm at 10 atm min^{-1}, then ramped to 400 atm at 5 atm min^{-1}.

can extend the range up to 14 EO units. Previous published work with SFC has shown EOs of up to 20 units of ethoxylation. Figure 5.3 shows C_{12} and C_{14} linear alcohols that have been ethoxylated. The units of ethoxylation are well separated up to almost 20 EO units. Geissler (1989) has published a comprehensive study showing the quantitation of ethoxylated alcohols, and Johnson *et al.* (1990) have shown how to determine relative ethoxylation rate constants by SFC. Other published work has shown non-ionic surfactants of different alcohols and chain lengths.

As with polyglycols, the use of SFC has progressed to the point of being able to provide quality-control determinations of ethoxylated alcohols. Figure 5.4 shows two runs on two different lots of an ethoxylated surfactant. Because of the extent of ethoxylation, GC could not show a difference between the two lots, and without adding chromophores HPLC could not be used. In the two chromatograms, identical conditions were used for

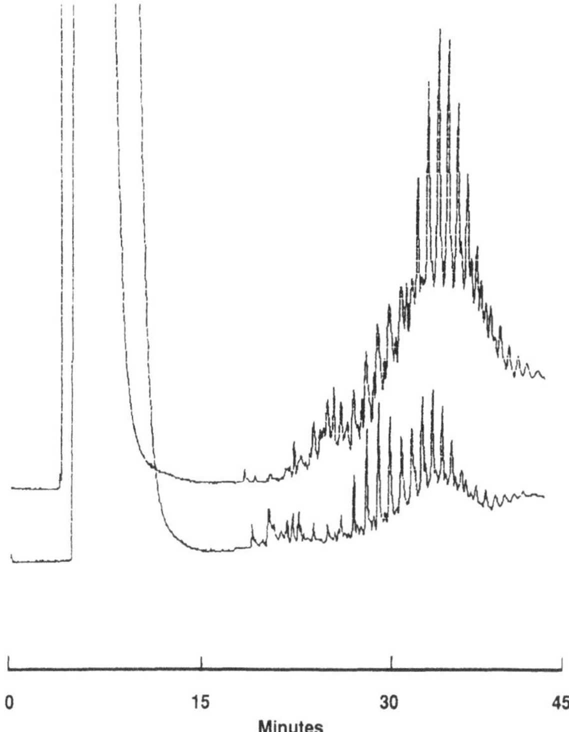

Figure 5.4 SFC comparison of ethoxylated surfactants. Conditions: CO_2 at 100 °C; FID at 350 °C; column: 10 m × 50 μm i.d. SB-Biphenyl-30; timed-split injection of 0.075 s; pressure program: 100 atm held for 10 min, then ramped to 400 atm at 10 atm min^{-1}.

the two lots. The top chromatogram represents the lot of material that was considered 'bad'. Visually, the two chromatograms show glaring differences in the EO distributions.

5.3.4 *Polystyrenes*

Polystyrenes were one of the first polymer classes studied by SFC. Klesper and Hartmann (1977) showed packed-column applications on polystyrenes, and Fjeldsted *et al.* (1983) showed the first capillary SFC analysis for polystyrenes. Hirata and Nakata (1984) have also shown packed capillary SFC applications for polystyrenes. Polystyrenes are used as insulating material and are prevalent in packaging materials and in the manufacture of foams. While most of the polystyrene work has been done using CO_2 as the mobile phase, Fujimoto *et al.* (1989) have used supercritical methylene chloride to elute polystyrenes of up to 2 900 000 molecular weight.

5.3.5 *Polysulfides*

Polysulfides are present in coating materials, petroleum products and also in rubber. Because of their high molecular weights and non-chromophoric nature they are very well suited for SFC analysis. Figure 5.5 shows a polysulfide of six oligomeric units. Because of the solvating power of supercritical fluids, component resolution within each series is possible. Polysulfides have also been analyzed using selective detectors. For these polymers a sulfur chemiluminescence detector (SCD) or a flame photometric detector (FPD) can be used. Pekay and Olesik (1989) have worked on polysulfides with SFC using an FPD.

5.3.6 *Polysiloxanes*

Depending on the molecular weights of polysiloxanes, they can be used as coatings, sealants, finishes, lubricants, insulators, and surfactants. Super-

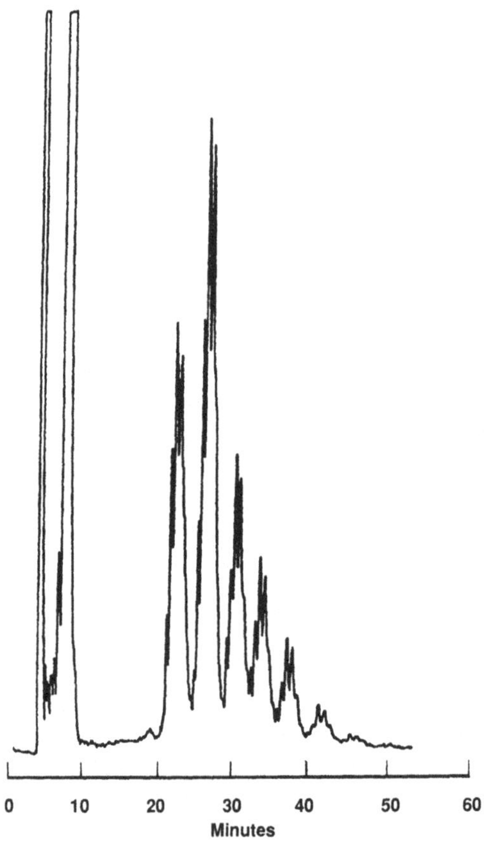

Figure 5.5 Polysulfide. Conditions: CO_2 at 150 °C; FID at 350 °C; split injection with split ratio of approx. 20:1; column: 5 m × 50 μm i.d. SB-Methyl-100; pressure program: 100 atm held for 10 min, then ramped to 400 atm at 10 atm min^{-1}.

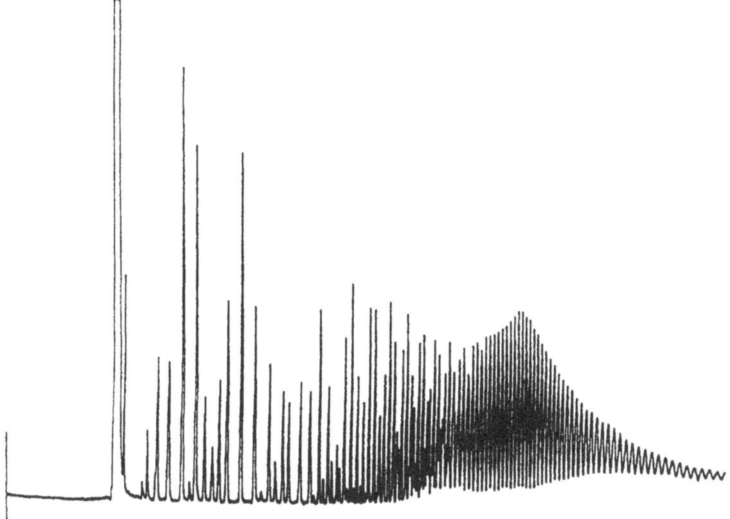

Figure 5.6 Silicone oil. Conditions: FID at 350 °C; mobile phase: CO_2; column: 5 m × 50 μm i.d. SB-Methyl-100; timed-split injection; simultaneous temperature and density programming. Courtesy of Nathan Porter, Dionex.

critical fluid technology has increased the range over which these polymers can be analyzed. HTGC has been used to achieve separation of up to 55 oligomeric units. With SFC, however, the detection range is much increased. Figure 5.6 shows a silicone oil that was analyzed using capillary SFC. This chromatogram shows baseline separation of 100 oligomeric units. Packed-column SFC also provides much more information for these silicone oils than does HTGC, but due to pressure drop constraints packed columns are not able to resolve oligomers as high as capillary columns. This run also points out the value that simultaneous temperature and density programming can have in increasing the oligomeric range and separation of many polymers. Most of the work in SFC utilizes solubility. However, there is also a volatility effect in SFC. There are many reports in which changing the temperature along with changing the pressure or density have shown advantageous effects on the chromatography. In this case, without the temperature programming, oligomers up to only about 60 units could be separated. Later *et al.* (1988*b*) have shown other examples in which simultaneous temperature and density programming have expanded the oligomeric detection range of polysiloxanes. Wenclawiak (1988) has also used negative temperature programming to optimize polymer separations.

5.3.7 *Polymer additives*

There have been many papers published using SFC for analysis of polymer additives. Raynor *et al.* (1988) have shown a series of 21 additives using

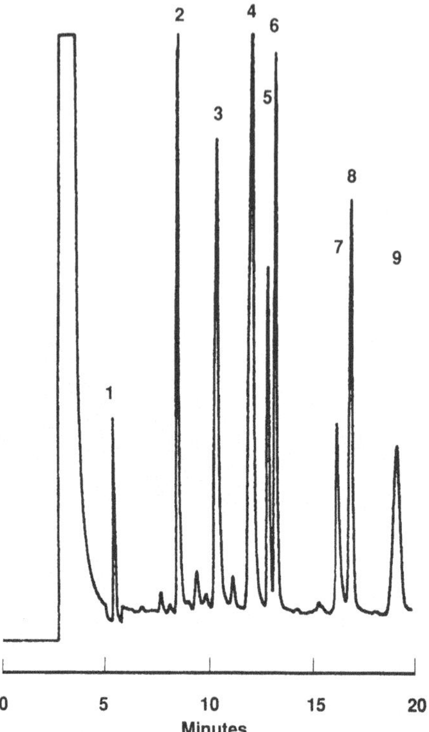

Figure 5.7 Polymer additives. Conditions: CO_2 at 75 °C; FID at 400 °C; timed-split injection at 0.075 s; column: 10 m × 50 μm i.d. SB-Biphenyl-30; density program: 0.25 g ml^{-1} held for 5 min, then ramped to 0.8 g ml^{-1} at 0.02 g ml^{-1} min^{-1}. Peak identification: (1) BHT 2; (2) Isonox 129; (3) Kenamide 0; (4) Kenamide E; (5) DLTDP (dilauryl thiodiproprionate); (6) Irganox 1076; (7) Santonox; (8) DSTDP (distearyl thiodiproprionate); (9) Irganox 1010.

FTIR and FID for detection. Many polymer additives can be analyzed by HPLC; however, in many cases, the sensitivity of the UV detector is not adequate. Polymer additives are used for a wide variety of applications. They can be UV stabilizers, mold release agents, antioxidants and binding agents. They are found in food products, surfactants, polymers, plastics, rubbers, cosmetics and textiles. Figure 5.7 shows the separation of nine commonly used polymer additives as found in an olefin stream. Some of these can be analyzed by GC, others by HPLC. SFC is the only technique that can separate all nine additives. Some other matrices that have been investigated in analyzing polymer additives are polypropylene films (Cotton *et al.*, 1991; Moulder *et al.*, 1989), and plastics (Arpino *et al.*, 1990).

5.3.8 Miscellaneous polymers

Besides those mentioned previously, numerous other examples of polymer analysis by SFC have been published. Selected examples of these are:

alkylene oxide–fatty alcohols (Onuska and Terry, 1988); fluorinated alkoxylates (White and Houck, 1986); lubricating oil additives (Ashraf *et al.*, 1991); poly(ethylene terephthalate) (Bartle *et al.*, 1991); and vinylnaphthalene oligomers (Schmitz *et al.*, 1988).

5.4 Supercritical fluid extraction

5.4.1 Off-line versus on-line SFE

Supercritical fluid extraction (SFE) has recently been receiving a great deal of attention as an alternative method for sample preparation. This attention has spread into the polymer industry. The advantages of using supercritical fluids to separate polymer additives, monomers and low-molecular-weight oligomers from high-molecular-weight polymers, and to remove polymeric materials from many finished products, are rapidly being realized and utilized in industry.

SFE offers several advantages to the polymer industry over other more conventional sample-preparation techniques such as liquid extraction, purge and trap, or digestion techniques. One important advantage of SFE is that it reduces the use of organic solvents in the laboratory. Typically Soxhlet or liquid extractions and distillations require the use of very large amounts of organic solvents. The initial costs of organic solvents are significant. However, the real expense of using organic solvents is often realized when the solvent is disposed of. Disposal costs for solvents are a rapidly increasing expense which SFE can greatly reduce. Also, regulatory agencies are beginning to require a reduction in the use of many organic solvents historically used for extraction purposes. This is due to their detrimental effects on both the environment and human health. The most commonly used supercritical extraction fluid is carbon dioxide. Carbon dioxide is a non-toxic, non-flammable solvent which requires no special disposal techniques and limits the danger of exposure to laboratory workers. Carbon dioxide offers other advantages for extraction purposes which have added to its popularity as a supercritical extraction fluid. This includes its low critical parameters which allow thermally labile substances to be extracted at temperatures just above 31 °C, the critical temperature for carbon dioxide. Furthermore, carbon dioxide in a closed extraction system limits the amount of free oxygen in that system, which can prevent oxidation of reactive compounds.

An additional advantage of SFE is that it often offers a reduction in the time required for an extraction. Because the diffusion coefficients are higher in a supercritical fluid than in a liquid–liquid system, the sample can be exposed to a larger amount of extraction fluid, with a subsequent faster migration of solutes from the matrix, in a shorter amount of time with SFE than with liquid-solvent based extractions. Many extractions which require several hours to be completed by liquid solvents can be completed in 1 h or less by using SFE. Another advantage of SFE involves the ability to affect greatly the solubility properties of the supercritical extraction fluid by making small changes in the density of the fluid. Thus, by making adjustments in the temperature or pressure of the supercritical fluid, it is possible to use a single SFE fluid to perform selective extraction of a wide range of analytes from an extensive range of matrices.

There are basically two different techniques for supercritical fluid extraction. The first approach is on-line extraction or coupled extraction. In on-line extraction the extraction cell is coupled directly to the chromatograph system. An on-line extraction typically replaces the standard chromatographic injection technique and is therefore limited in many respects. First, on-line extraction limits the techniques by which the extracted material can be analyzed. Thus far, on-line SFE has been limited to being coupled with gas chromatography, liquid chromatography, or supercritical fluid chromatography. Generally in on-line extraction, the entire amount of extracted material is collected and delivered to the chromatographic column, which results in the inability to analyze the extracts by any other analytical technique. Also, the sample size can be a limiting factor for on-line extraction. On-line SFE is inherently a very sensitive technique due to the fact that the entire amount of extracted material is usually transferred into the chromatographic system. A large sample with a high concentration of extractable analytes can lead to a loss of chromatographic resolution due to an overloading of the column. For capillary columns with internal diameters of 50 μm, an injected volume of over approximately 100 nl will cause a 1% loss in resolution, while for typical packed columns, which are 25 cm × 1 mm i.d., the injected volume must remain below 1 μl to avoid a significant loss of resolution (Peaden and Lee, 1983). So, the total amount of extracted material must be kept below these levels for on-line SFE. And finally, on-line extraction can be limiting in terms of sample throughput. Typically, on-line SFE is limited to a single cell extraction during which the entire chromatographic system is inoperative. Thus, one must complete the entire extraction and analytical process on a single sample before being able to advance to the next sample. Therefore, this technique can become too time-consuming for a large number of samples.

The second approach to supercritical fluid extraction is off-line SFE. Off-line SFE involves collecting the extracted analytes in a manner which is exclusive of any analytical technique. Off-line SFE allows analysis of extracted material by virtually any method, and thus overcomes many of the limitations associated with on-line SFE. The advantages of off-line SFE include the flexibility of being able to analyze the extracted material by a wide range of methods including gravimetric, spectroscopic, fluorescent, and chromatographic techniques, or any possible combination of these. For example, a common procedure is to extract a polymeric material by off-line SFE, collect the sample in a dry trapping region which contains no liquid solvent or dry trapping material, analyze the amount of material extracted gravimetrically, and then add solvent to the trapping vial so that the sample can be injected into a chromatographic system for separation and identification of the polymer or extracted additives. In addition, off-line SFE also offers the advantage of the ability to extract large sample quantities. Because the extracted analytes in off-line SFE are collected in a region that is independent of the analytical technique, the resultant extracts can be diluted to lower concentrations with solvent or, conversely, concentrated to higher concentrations. Thus, typical extraction cell sizes for off-line SFE range from approximately 0.5 ml to 50 ml volumes. Larger sample sizes offer a greater homogeneity and with larger samples it becomes easier to detect very low concentrations of analytes. For instance, if one wanted to extract a polymer additive which was present in a matrix at a concentration of 1 ppb, a 100% extraction efficiency of a 1 g sample would lead to only 1 ng of extracted material. If this extracted additive was collected in 1 ml of solvent or more, it would be very difficult to detect by many common analytical techniques. Yet, if one were able to increase the amount of material extracted to 20 g, which is feasible with off-line extraction, but often is too large for on-line extraction, a 100% efficient extraction would furnish 20 ng of material, an amount which is much simpler to detect by conventional methods of analysis. In addition, many of the commercially available off-line extraction instruments allow simultaneous extraction of more than one sample, which can also increase the amount of material that can be extracted in a given amount of time.

5.4.2 Method devlopment

The supercritical fluid extraction of polymers and polymer additives is a relatively simple technique, but optimization of the extraction requires decisions concerning the extraction fluid, the extraction pressure and temperature, and the techniques that are used to pack the extraction cell.

5.4.3 Extraction fluid

Selection of the correct supercritical extraction fluid is dependent upon both the polymer that is to be extracted and the matrix from which one is extracting it. As mentioned above, supercritical carbon dioxide is the most commonly used fluid. Carbon dioxide under supercritical conditions is a good choice for extraction of polymeric material with low polarity and low to moderate molecular weight from matrices that have little organic character. Most polymer additives and polymers with molecular weights up to approximately 10 000 daltons are soluble in CO_2. Polymers having molecular weights higher than 10 000 require a fluid other than CO_2. Propane and methylene chloride under supercritical conditions have proved to be useful supercritical fluids for the solvation of these higher-molecular-weight polymers.

Organic modifiers are occasionally added to the carbon dioxide for extraction of polymeric materials in order to enhance its extraction efficiency. These co-solvents are typically added into the carbon dioxide at levels of 10% or lower for SFE purposes. Modifiers can increase the solubility of the solutes in the supercritical fluid, or they can be added to overcome bonding effects between the analytes of interest and an organic matrix. Typically, the polymers and polymer additives are soluble in the carbon dioxide but an organic solvent is required to either swell the matrix or to break the strong physical or chemical bonds between the analytes of interest and the matrix. Thus, the organic modifier should be selected based upon both the solubility properties of the analytes which are to be extracted and the properties of the matrix.

5.4.4 Extraction pressure and temperature

Many analytes including polymeric materials exhibit solubility maxima in supercritical solvents. For most analytes there is a density at which the solubility is maximized. Several different models have been developed to predict the solubility of compounds in supercritical fluids. The first model that looked at the solubility of an analyte in a supercritical fluid was developed by Giddings *et al.* (1968) in which they introduced a term called 'threshold pressure'. The threshold pressure is the point at which the solute first becomes soluble and can be a useful term for determining the most efficient extraction pressure, particularly for SFE of polymers. More contemporary models include, but are not limited to, the model by Gitterman and Procaccia (1983), which predicts solubility changes versus pressure, and King's (1989) discussion on estimating fractionation pressures for SFE.

The most common means of determining the best extraction parameters for a given set of compounds is through experimentation. However, a basic understanding of the physical properties of the analytes that are to be extracted can greatly add to the ability to choose the optimal extraction conditions. If the melting point of the solute is known, the extraction should be performed at a temperature just above this point. Temperatures in this range aid in extraction efficiencies due to the breakdown of the forces binding the solute together and the increased diffusion rate of the solute through the supercritical fluid. Also, knowing the vapor pressure of the analytes can offer information about the optimum pressure to extract at, particularly for extractions in which fractionation is the objective. Experimentation has shown that, for certain samples, a very high extraction pressure is necessary to achieve acceptable extraction efficiencies. One example of this type of sample was recently presented by Richter (1991), in which three classes of compounds, polycyclic aromatic hydrocarbons (PAHs), polychlorinated biphenyls (PCBs) and chlorinated pesticides, were extracted from an oyster tissue sample. This presentation summarized the average recoveries of the different compounds by class based upon two different extraction pressures. At an extraction pressure of 6000 psi, the extraction recoveries for all of the compounds were significantly lower than at the higher 10 000 psi pressure.

5.4.5 Extraction cell packing techniques

The means by which the sample is placed into the extraction cell can also have a profound effect on the extraction efficiency obtained. Channeling of the CO_2 through the sample in the cell can affect the amount of sample that is exposed to the extracting fluid, which can lead to lower extraction efficiencies. Thus, extraction cells should be packed in a way that prevents the presence of cavities through which the supercritical fluid can move. The most common means of preventing chaneling is to ensure that the extraction cell is completely full of sample. However, this presents a possible difficulty with polymeric samples. When many polymeric materials are exposed to carbon dioxide under the high pressures required for supercritical fluid extraction, they expand. If the extraction cell is packed full of a polymer which swells, two possible problems can occur. First, the polymer may extrude from the cell during an extraction, and second, the extraction cell may be permanently 'locked' closed, due to the polymer expansion. Probably the best means for addressing the swelling effect for polymers while still preventing channeling effects is to fill the extraction cell to a level that is low enough to accommodate any swelling that may occur, and then to fill the rest of the extraction cell with a compressible support material such as clean glass wool.

5.5 Applications of on-line extraction

5.5.1 Polymer additives

Polymer additives have become routinely analyzed by SFC. Indeed, SFC is fast becoming the technique of choice for polymer additives. Since they have been well characterized by SFC, it follows that they should be one of the major applications that have been studied by SFE. Several publications have shown quantitative results for on-line SFE–SFC from a polyethylene

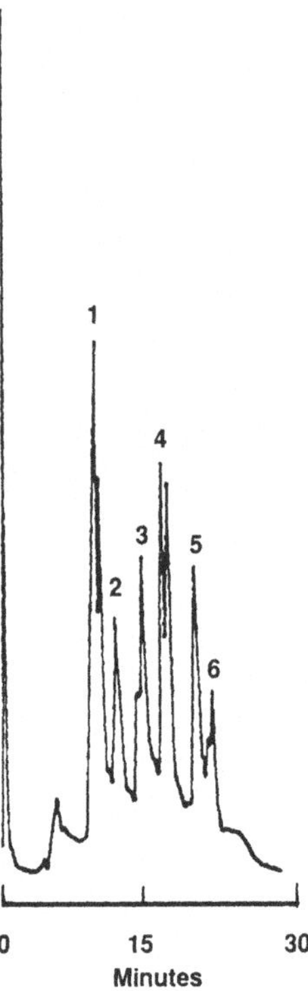

Figure 5.8 On-line SFE–SFC of polymer additives from polypropylene pellets. Extraction conditions: CO_2 at 75 °C; pressure: 250 atm; extraction time: 5 min; extraction weight: 5 mg; chromatography conditions: CO_2 at 125 °C; FID at 400 °C; column: 5 m × 50 μm i.d. SB-Methyl-100; pressure program: 125 atm ramped to 400 atm at 20 atm min^{-1}; additives at 100 ppm levels. Peak identification: (1) Tinuvin P; (2) Isonox 129; (3) Kenamide 0; (4) Kenamide E, Weston 626; (5) DLTDP; (6) DSTDP.

matrix. Ashraf-Khorassani and Levy (1990) have published work using on-line extraction coupled to packed-column SFC of polymer additives extracted from low-density polyethylene (LDPE), while Cotton *et al.* (1991) used on-line SFE coupled to capillary SFC to analyze quantitatively several polymer additives from various matrices such as polypropylene, nylon, and poly(ethylene terephthalate). In an earlier section in this chapter, polymer chromatography of additives from an olefin stream was described (Figure 5.7). Figure 5.8 shows an on-line extraction of the final product, polypropylene pellets, to see if any of the additives are still detectable. Extracting the polypropylene pellets for 5 min and cryofocusing the extract allow the detection of 100 ppm of each of the six polymer additives. Combining the extraction and chromatography run times still allows analysis of the additives in less than 25 min. A similar sample using Soxhlet extraction followed by either HPLC or GC would take over 48 h. In this run no liquid solvent is used. Table 5.1 shows the type of reproducibility that is possible with on-line extraction of this sample. Three different

Table 5.1 Reproducibility of on-line SFE for polymer additives at 100 ppm extracted from polypropylene pellets

Additive	1	2	3	RSD(%)
Tinuvin P	15448504	15205475	15501321	2.69
Isonox	9807312	10184529	10563160	3.71
Kenamide 0	11415963	11410160	11397882	1.61
DLTDP	7476366	7018918	7715091	4.78
DSTDP	7022502	6793291	6470220	4.10

extractions under identical conditions were made. Each extract was cryogenically focused directly on an SFC column and then analyzed by SFC. Note that the RSD percentages for the three extractions are a function of both the extraction efficiency and the chromatographic RSDs.

5.5.2 *Hydrocarbon waxes*

It has been just a few years since the first report by Richter *et al.* (1986) on on-line extraction appeared. In this short time, methods have been developed using on-line SFE for quality control. Figure 5.9 is just one

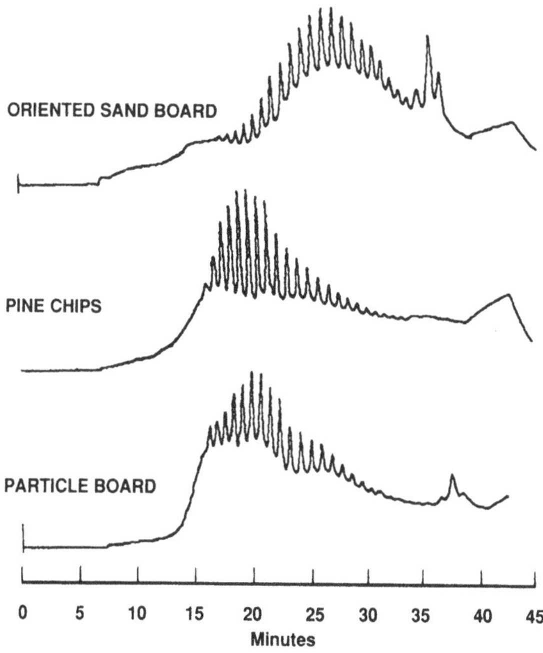

Figure 5.9 Comparison of waxes from wood pulp by on-line SFE–SFC. Extraction conditions: CO_2 at 50 °C; pressure: 300 atm; extraction time: 5 min; chromatography conditions: mobile phase: CO_2; FID at 350 °C; simultaneous temperature and pressure program; column: 3 m × 50 μm i.d. SB-Octyl-50.

example of this. In the paper industry it is necessary to check for the amount of wax on various paper products. It is important to be able to both detect and quantitate the amount of wax that has been applied to the surface of wood pulp. Three different wood pulps (sand board, pine chips, and particle board) were extracted on-line to a capillary SFC to check their corresponding wax distribution. The same method was used for all three pulps. While there is a different wax distribution in each sample, the molecular-weight range of the waxes is 300–400 daltons. Quantitation of these results allows the relative amount of each wax bound on the individual wood pulp to be determined.

5.5.3 Polystyrene

Polystyrene is used for, among other things, the making of drinking cups. Within polystyrene cups are various components. Among these, Polywax 1000 is used to add a gloss to the polystyrene cup. Other components may include silicon dioxide, various additives, zinc stearate, and other waxes. Figure 5.10 is an on-line extraction of a piece of a polystyrene cup. In this cup, Polywax 1000 is present in concentrations of between 100 and 1000 ppm. Retention times for standards were used to identify various components of the polystyrene cup. The extraction and chromatography were not optimized for all the components present. Figure 5.11 is an on-line

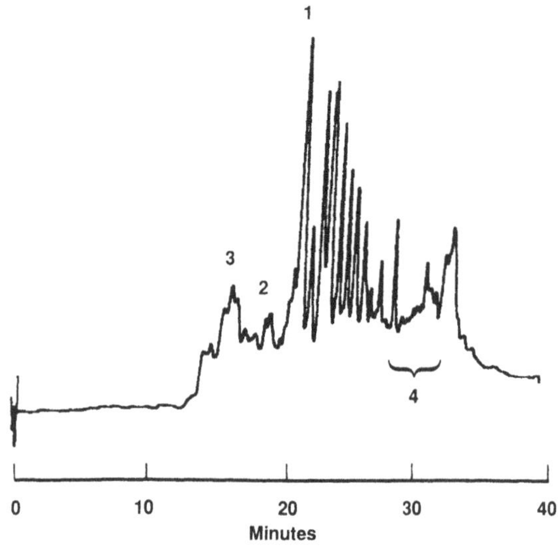

Figure 5.10 On-line SFE–SFC of polystyrene T-113. Extraction conditions: CO_2 at 75 °C; pressure: 400 atm; extraction time: 10 min; chromatography conditions: mobile phase: CO_2 at 100 °C; FID at 400 °C; column: 10 m × 50 μm i.d. SB-Octyl-50; pressure program: 100 atm ramped to 400 atm at 10 atm min^{-1}. Peak identification: (1) Carbowax 400 (major peak); (2) Triton X-102; (3) zinc stearate; (4) Polywax 1000.

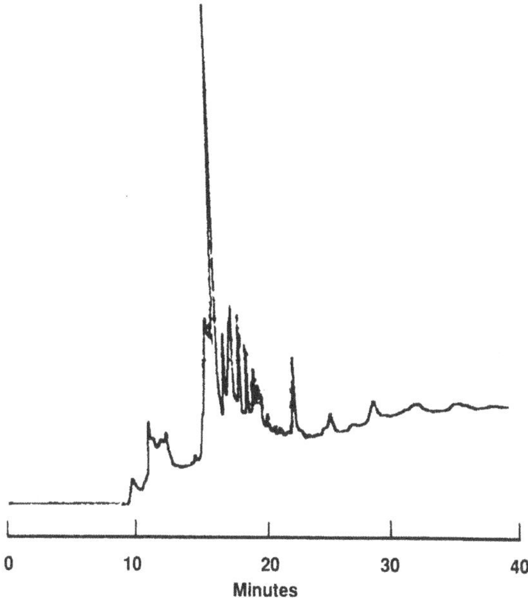

Figure 5.11 Polystyrene without Polywax 1000. Conditions were the same as in Figure 5.10.

extraction of a different polystyrene cup that did not contain any of the Polywax 1000. The analysis was done using the same method as in Figure 5.10. On-line SFE–SFC has become a powerful technique for the analysis of low levels of waxes and/or additives.

5.6 Applications of off-line extraction

5.6.1 Polyglycols

Polyglycols are popular industrial chemicals. They have a diverse range of uses which is often dependent upon their molecular weights. Polyglycols are frequently added into other materials to be used as plasticizers, binders, softeners, and surface coatings. Because many polyglycols are readily soluble in supercritical carbon dioxide, SFE provides an excellent means for extracting polyglycols from the various components to which they are added. One example of the extraction of a polyglycol material is shown in Figure 5.12. This figure demonstrates the SFC results from the supercritical fluid extraction of a Carbowax blend from a cigarette filter. Carbowax is sometimes added into cigarette filters in order to retard the flow of smoke and other cigarette components through the filter. In order to extract the filter, the paper coating was removed from around the outside of the filter. Then the entire filter was simply placed into the

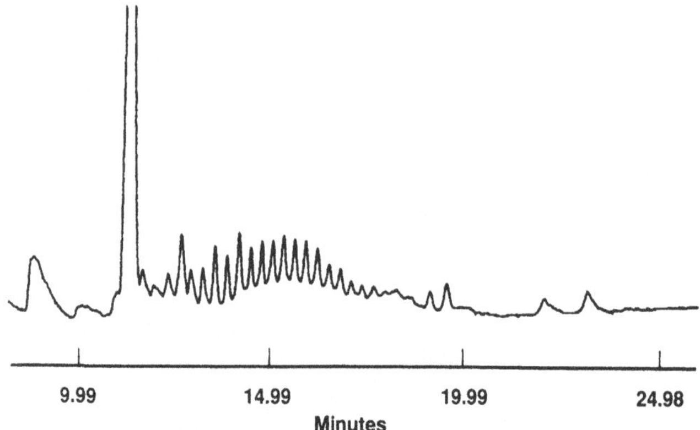

Figure 5.12 SFC chromatogram of Carbowax extracted off-line from cigarettes. Extraction conditions (off-line): CO_2 at 75 °C; pressure: 300 atm; restrictors: 150 °C; extraction time: 30 min; SFC conditions: CO_2 at 75 °C; FID at 350 °C; timed-split at 0.05 s; column: 5 m × 50 μm i.d. SB-Methyl-100; multi-linear density program.

extraction cell without any further preparation. The extraction was performed using 100% CO_2 as the extraction fluid. The extraction conditions were 300 atm pressure and 75 °C for a total of 30 min. Because of the molecular weight of the extracted material, this extraction required that the restrictor region leading to the collection vials was kept at a temperature of 150 °C. If the restrictors had not been heated, plugging of the restrictors would have stopped the flow of extraction fluid through the system and ended the extraction process in a very short time. The extracted material was collected in methylene chloride and taken to a supercritical fluid chromatograph for analysis. By extracting with supercritical fluid technology, both the amount of solvent used and the analysis time required were greatly reduced from those required in the conventional methodology.

5.6.2 *Ethoxylated alcohols/surfactants*

Ethoxylated alcohols are often added to the surface of materials as non-ionic surfactants. These surfactants reduce the surface tension between the material and a liquid. Supercritical fluid chromatography has proven to be an excellent means of analyzing these materials; thus SFE should be a viable technique for separating these compounds from the surface of the materials to which they are added. The chromatogram in Figure 5.13 demonstrates the feasibility of this application. This figure shows the results obtained for the SFE of a non-ionic surfactant blend from a facial tissue paper sample. The surfactant was very soluble in CO_2 and thus did not require any organic modifier for complete extraction. The sample was

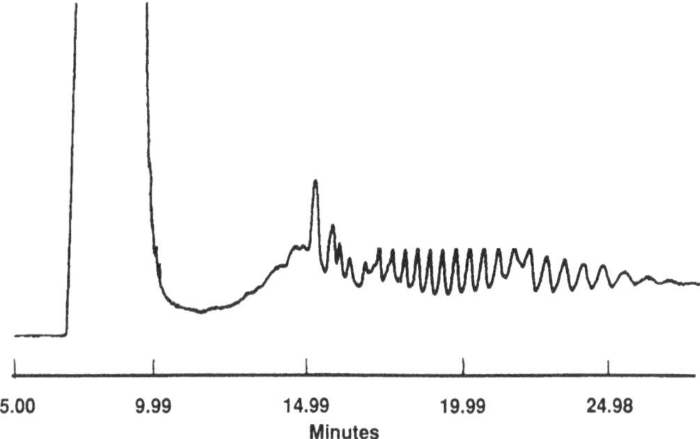

Figure 5.13 SFC of surfactant extracted off-line from tissue paper. Extraction conditions (off-line): CO_2 at 75 °C; pressure: 275 atm; restrictors: 150 °C; extraction time: 30 min; SFC Conditions: CO_2 at 100 °C; FID at 350 °C; timed-split at 0.100 s; column: 10 m × 50 μm i.d. SB-Methyl-100; pressure program: 100 atm held for 5 min, then ramped at 20 atm min^{-1} to 250 atm, then to 300 atm at 2 atm min^{-1}, then to 400 atm at 50 atm min^{-1}.

extracted with SFE-grade CO_2 at 275 atm and 75 °C. The entire surfactant blend was removed from approximately 0.25 g of paper in 30 min. Because paper is a dry matrix which has proven to work very well with SFE, no additional sample preparation was required prior to the extraction. The samples were simply weighed and placed into the extraction cells. The extraction was followed by analysis with capillary SFE, which provided excellent resolution of the components of the extracted surfactant.

5.7 Conclusions

In this chapter, data have been presented showing polymer applications where SFC and SFE are presently being used to solve many difficult application problems. As the use of supercritical fluids, for both extraction and chromatography, continues to grow, new and better solutions to current problems facing the polymer industry will be developed.

References

Arpino, P.J., Dillettato, D., Nguyen, K. and Bruchet, A., (1990), Investigation of antioxidants and UV stabilizers from plastics. Part 1: Comparison of HPLC and SFC; Preliminary SFC/MS study. *J. High Resolut. Chromatogr.*, **13**, 5–12.

Ashraf, S., Bartle, K.D., Clifford, A.A. and Moulder, R., (1991), Trace analysis of agrochemicals by SFC. *J. High Resolut. Chromatogr.*, **14**, 29–32.

Ashraf-Khorassani, M. and Levy, J.M., (1990), Quantitative analysis of polymer additives in low density polyethylene using SFE/SFC. *J. High Resolut. Chromatogr.*, **13**, 742–747.

Bartle, K.D., Boddington, T., Clifford, A.A., Cotton, N.J. and Dowle, C.J., (1991), Supercritical fluid extraction and chromatography for the determination of oligomers in poly(ethylene terephthalate) film. *Anal. Chem.*, **63**, 2371–2377.

Campbell, C., (1991), Some applications of SFC to petroleum additive development. *Federation of Analytical Chemistry and Spectroscopy Societies*, 6–11 October, 1991, Anaheim, CA, USA, paper 109.

Chester, T.L., Bowling, D.J., Innis, D.P. and Pinkston, J.D., (1990), Capillary supercritical fluid chromatography at pressures above 400 atmospheres. *Anal. Chem.*, **62**, 1299–1301.

Cotton, N.J., Bartle, J.D., Clifford, A.A., Ashraf, S., Moulder, R. and Dowle, C.J., (1991), Analysis of low molecular weight constituents of polypropylene and other polymeric materials using on-line SFE–SFC. *J. High Resolut. Chromatogr.*, **14**, 164–168.

Fjeldsted, J.C., Jackson, W.P., Peaden, P.A. and Lee, M.L., (1983), Density programming in capillary SFC. *J. Chromatogr. Sci.*, **21**, 222–225.

Fujimoto, C., Watanabe, T. and Jinno, K., (1989), Size exclusion chromatography of polystyrenes with supercritical dichloromethane. *J. Chromatogr. Sci.*, **27**, 325–328.

Geissler, P.R., (1989), Quantitative analysis of ethoxylated alcohols by SFC. *J. Amer. Oil Chem. Soc.*, **66**, 685–689.

Giddings, J.C., Myers, M.N., McLaren, L. and Keller, R.A., (1968), High pressure gas chromatography of nonvolatile species. *Science*, **162**, 67–73.

Gitterman, M. and Procaccia, I., (1983), Quantitative theory of solubility in supercritical fluids. *J. Chem. Phys.*, **78**, 2648–2654.

Harvey, M.C. and Stearns, S.P., (1984), High-speed switching of liquid chromatographic injection valves. *Anal. Chem.*, **56**, 837–839.

Hirata, Y. and Nakata, F., (1984), SFC with fused-silica packed columns. *J. Chromatogr.*, **295**, 315–320.

Johnson, A.E., Jr., Geissler, P.R. and Talley, L.D., (1990), Determination of relative ethoxylation rate constants from SFC analysis of ethoxylated alcohols. *J. Amer. Oil Chem. Soc.*, **67**, 123–131.

King, J.W., (1989), Fundamentals and applications of supercritical fluid extraction in chromatographic science. *J. Chromatogr. Sci.*, **27**, 355–364.

Klesper, E. and Hartmann, W., (1977), Parameters in supercritical fluid chromatography of styrene oligomers. *Polymer Lett.*, **15**, 707–712.

Knowles, D.E., Nixon, L., Campbell, E.R., Later, D.W. and Richter, B.E., (1988), Industrial applications of SFC: Polymer analysis. *Fresenius Z. Anal. Chem.*, **330**, 225–228.

Later, D.W., Bornhop, D.J., Lee, E.D., Henion, J.D. and Wieboldt, R.C., (1988*a*), Detection techniques for capillary SFC. *Liq. Chromatogr. Gas Chromatogr.*, **5**, 804–816.

Later, D.W., Campbell, E.R. and Richter, B.E., (1988*b*), Synchronised temperature/density programming in capillary supercritical fluid chromatography. *J. High Resolut. Chromatogr.*, **11**, 65–69.

Lee, M.L. and Markides, K.E. (Eds), (1990), *Analytical Supercritical Fluid Chromatography and Extraction*, Chromatography Conferences, Inc., Provo, UT, USA.

Moulder, R., Kithinji, J.P., Raynor, M.W., Bartle, K.D. and Clifford, A.A., (1989), Analysis of chemical additives in polypropylene films using capillary SFC. *J. High Resolut. Chromatogr.*, **12**, 688–691.

Onuska, F.J. and Terry, K.A., (1988), SFC of alkylene oxide–fatty alcohol condensates: their quantification in water samples. *J. High Resolut. Chromatogr.*, **11**, 874–877.

Peaden, P.A. and Lee, M.L., (1983), Theoretical treatment of resolving power in open tubular column SFC. *J. Chromatogr.*, **259**, 1–16.

Pekay, L.A. and Olesik, S.V., (1989), SFC/FPD: determination of high molecular weight compounds. *Anal. Chem.*, **61**, 2616–2624.

Raynor, M.W., Bartle, K.D., Davies, I.L., Williams, A., Clifford, A.A., Chalmers, J.M. and Cook, B.W., (1988), Polymer additive characterization by capillary SFC/FTIR microspectrometry. *Anal. Chem.*, **60**, 427–433.

Richter, B.E., (1989), *Molecular Weight Determination*, Cooper, A.R. (Ed.). Wiley Interscience, New York, pp. 373–390.

Richter, B.E., (1991), *National Meeting Assoc. Off. Anal. Chem.*, Phoenix, AR, USA, 12–15 August, 1991.

Richter, B.E., Knowles, D.E., Andersen, M.R. and Later, D.W., (1986), *Eastern Analytical Symposium*, New York, NY, USA, paper 102.

Richter, B.E., Knowles, D.E., Andersen, M.R., Porter, N.L., Campbell, E.R. and Later, D.W., (1988), Reproducibility in capillary SFC: Comparison of injection techniques. *J. High Resolut. Chromatogr.*, **11**, 1–5.

Schmitz, F.P., Gemmel, B. and Leyendecker, D., (1988), Separation of 2-vinylnaphthalene oligomers by open-tubular capillary SFC. *J. High Resolut. Chromatogr.*, **11**, 339–340.

Sie, S.T. and Rijnders, G.W.A., (1967), High pressure GC and chromatography with supercritical fluids, II. Permeability and efficiency of packed columns with high-pressure gases as mobile fluids under conditions of incipient turbulence. *Sep. Sci.*, **2**, 699–727.

Wenclawiak, B.W., (1988), Negative temperature gradients for density programming in SFC. *Fresenius Z. Anal. Chem.*, **330**, 218–221.

White, C.M. and Houck, R.K. (1986), SFC and some of its applications: A review. *J. High Resolut. Chromatogr.*, **9**, 4–17.

6 Applications of supercritical fluids in food science

J.R. DEAN

6.1 Chemical derivatization

The choice of carbon dioxide as the supercritical fluid for chromatography and extraction is often concerned with its low cost, non-toxicity, inertness and availability in high purity. While these requirements are indeed valid, the range of analytes to which supercritical carbon dioxide can be applied is limited to non-polar–moderately polar analytes. While this may be acceptable to demonstrate the effectiveness of supercritical fluid chromatography (SFC) and/or supercritical fluid extraction (SFE), frequently it can provide a major limitation for real samples such as foodstuffs. Initial investigations of the use of modified carbon dioxide have met with some success. This typically involves the addition, via a second pump or a premixed cylinder, of an organic modifier (entrainer), often methanol. The addition of the modifier acts to increase the solvent strength of the supercritical carbon dioxide and to act as a competitor with adsorbed analytes for active sites on the matrix (Hills *et al.*, 1991). An alternative approach to facilitate extraction and/or chromatography of polar analytes involves chemical derivatization. This may have the effect of reducing the polarity of the target analyte while increasing its solubility in supercritical carbon dioxide. In chromatography this would have the effect of shortening the retention time and improving peak tailing, whereas in extraction the additional benefit of chemically derivatizing the active sites to prevent resorption of the analyte is possible. Recently, several publications have appeared which discuss the feasibility of utilizing this method of derivatization in SFC and SFE. Cole *et al.* (1991) undertook a preliminary investigation of common gas chromatography (GC) silylating agents to determine their utility for derivatizing a variety of polar functional groups and for improving peak shapes in capillary SFC. Of particular note is the potential to block polar functional groups more efficiently by the use of bulkier derivatizing agents. This was demonstrated most effectively by the sharp, single peak obtained when octadecanoic acid was derivatized with *N*-methyl-*N*-(*tert*-butyldimethylsilyl)-trifluoroacetamide to produce the *tert*-butyldimethylsilyl derivative. Although this work was not directly in the food science area the potential to extrapolate the method is a viable proposition.

Alternative approaches for chemical derivatization have focused on the ability to extract polar analytes. The use of the commercially available silylating agent Tri-sil concentrate, a mixture of hexamethyldisilane and trimethylchlorosilane, for the further extraction of analytes that had previously been exhaustively extracted using supercritical carbon dioxide, was demonstrated (Hills *et al.*, 1991). This method was applied to roasted coffee beans and Japanese tea leaves. The effectiveness of this method is illustrated in Figure 6.1 in which gas chromatograms of extracts from roasted coffee beans with and without the additional of the silylating agent are shown. Figure 6.1(a) shows the extract obtained when supercritical

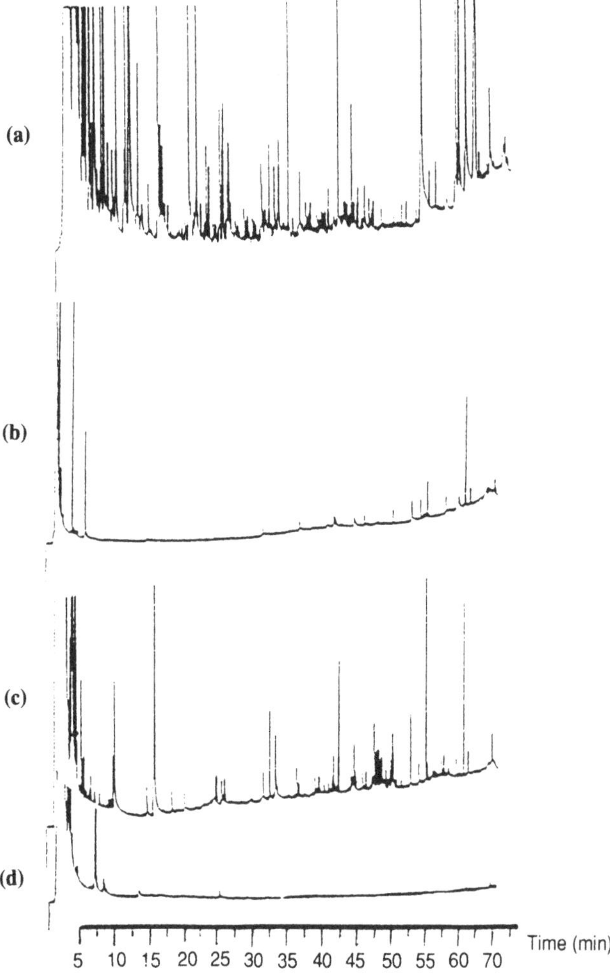

Figure 6.1 Gas chromatography–flame ionization detection of extracts from roasted coffee beans (10 min static extraction at 200 atm and 80 °C; extract collected in 1 ml of acetone). For reference to parts (a)–(c) see text. Chromatogram (d) is a procedural blank. Reprinted with permission from Hills *et al.* (1991). Copyright © 1991 American Chemical Society.

carbon dioxide at 200 atm and 80 °C was used in a static extraction for 10 min, followed by dynamic extraction and analyte collection in 1 ml of acetone. While these conditions were not sufficient to extract all the analytes present, the chromatogram in Figure 6.1(b) shows the situation after five repeat extractions on the same roasted coffee bean sample. To this exhaustively extracted sample was added 1 ml of the Tri-sil concentrate reagent and the extraction was repeated a sixth time (Figure 6.1(c)). As can be observed from Figure 6.1(c) additional analytes have been extracted. Identification of some of the peaks by gas chromatography–mass spectometry (GC–MS) showed that chemical derivatization had in fact occurred and liberated trimethylsilyl derivatives of a range of compounds. Perhaps more surprising was the additional extraction of underivatized compounds, a feature attributed to the presence of the derivatizing agent.

A derivatization approach with the same goal in mind has been the use of an ion-pair methylating reagent, trimethylphenylammonium hydroxide (Hawthorne *et al.*, 1992). This particular ion-pairing reagent was chosen because of its solubility in methanol. Quantitative recoveries of >90% were obtained for the derivatization and extraction of native and spiked acid herbicides, 2,4-dichlorophenoxyacetic acid and Dicamba, from soil and sediment. The *in situ* derivatization of polar analytes by either silylation or ion-pairing reagent will aid the applicability of supercritical fluid extraction and chromatography in food science. While continued work is obviously necessary, these preliminary papers provide an undoubted major step forward.

6.2 Solid-phase extraction

Most of the published material on SFE is concerned with the extraction of analytes from solid or semi-solid matrices. While this is obviously beneficial and necessary for a large number of samples in food science, it obviates the extraction of analytes from aqueous samples. To date, relatively few papers that investigate the extraction of analytes from aqueous matrices have appeared (Hedrick and Taylor, 1989, 1990). Taylor and co-workers achieved extraction from aqueous samples by directly passing supercritical carbon dioxide through the solution. Alternatively, the use of a combined system using solid-phase extraction (SPE) and SFE has been used by several groups (Kane *et al.*, 1991; Hawthorne *et al.*, 1992; Liu *et al.*, 1992). This involves adsorbing the analyte from its aqueous matrix on to either octadecylsilane cartridges (Liu *et al.*, 1992) or Empore disks (Kane *et al.*, 1991; Hawthorne *et al.*, 1992) followed by SFE. This method has been applied to non-ionic surfactants in water (Kane *et al.*, 1991), wood soot leachate (Hawthorne *et al.*, 1992) and the drug metabolite, mebeverine alcohol, from plasma (Liu *et al.*, 1992).

6.3 Multidimensional chromatography

The complex nature of food samples frequently requires the use of sample clean-up/pretreatment prior to a single chromatographic run. This has been demonstrated using supercritical fluids in which both extraction and chromatography have been involved prior to MS detection (Ramsey *et al.*, 1989). In this paper a tandem MS arrangement was used. This enabled the mass spectrometer to purify further the sample (fish tissue) prior to detection in the quadrupole mass spectrometer. While this may be possible, the practical skills required, the reliability and the high cost of the instrumentation make this method unsatisfactory for routine analysis. The use of multidimensional chromatography provides a viable alternative. In such a system (Moulder *et al.*, 1991), fractions of the eluate from a packed capillary size-exclusion column (10 μm Spherisorb silica particles) were selectively transferred via a solvent-vented interface to an open tubular capillary SFC (50% cyanopropyl-substituted methylpolysiloxane, 0.25 μm film, fused-silica) column. The system was used to investigate the migration of additives from polymeric food wrappings. An approved method involves a measure of the increase in mass of olive oil following prolonged contact with the wrapping material. An improved method may involve chromatographic analysis. The feasibility of the SEC–SFC approach was investigated using a 0.5% solution of olive oil in DCM containing two additives, Irganox 1330 and Irgafos 168. Figure 6.2 highlights the response from the combined chromatographic coupling. A 6.6 μl fraction was transferred from the SEC column to the SFC. The SFC trace showed minimal interference from any co-transferred oil components.

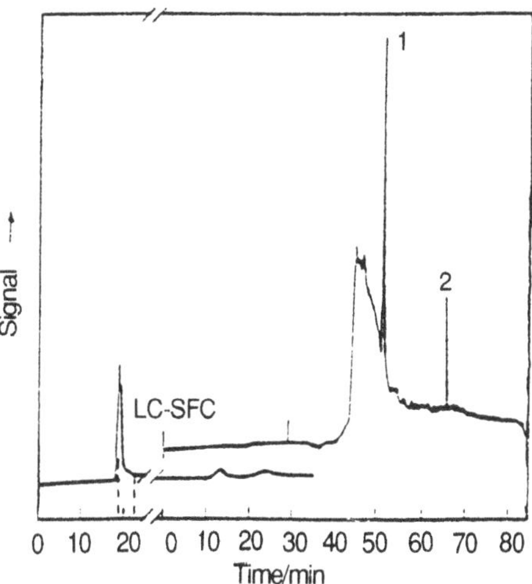

Figure 6.2 Size exclusion chromatography–supercritical fluid chromatography of an olive oil–polymer additive mixture. SEC column: 350 mm × 0.32 mm i.d.; 10 μm Spherisorb silica particles; 3 nm pore size; mobile phase, DCM; flow rate 2 μl min^{-1}. Peak identification: (1) Irgafos 168; (2) Irganox 1330. Reproduced with permission from Moulder *et al.* (1991).

6.4 Drug and chemical residues

Of particular concern in food science is the occurrence of drug and chemical residues in foodstuffs. These substances are introduced into the food chain from procedures applied by man to plants or animals which then become our food. Monitoring of their occurrence and concentration then becomes critical to any regulatory agency.

6.4.1 *Dioxins*

Dioxins and dibenzofurans are fat-soluble toxic compounds that accumulate in, for example, fish liver and milk. In a preliminary investigation, Jakobsson *et al.* (1991) described a method to extract dioxins from cod liver oil using supercritical carbon dioxide with and without an ethanol modifier. The preliminary results show that it is possible to extract dioxins from a de-acidified cod liver oil at pressures of 150 bar or less. However, while the use of the modifier improved the extraction of free fatty acids from the non-polar triglyceride fraction, no noticeable improvement in dioxin extraction was noted.

6.4.2 *Pesticides*

The US Department of Agriculture's Food Physical Chemistry Research Section has investigated the use of supercritical carbon dioxide as an alternative sample clean-up technique for organochlorine pesticides from fats (France *et al.*, 1991*a*). The replacement of clean-up techniques that require organic solvents would extend the beneficial aspects of SFE to the clean-up steps. This would reduce the organic solvent use and its attendant waste-disposal problems in both the extraction and the clean-up procedures. The supercritical fluid clean-up procedure uses either an alumina or a silica column for the separation of organochlorine residues from chicken fat and lard. Results were presented from chicken fat with incurred residues of heptachlor epoxide, dieldrin and endrin, and lard spiked with lindane, heptachlor, heptachlor epoxide, dieldrin, endrin and *o,p'*-DDT. The same group reported enhanced supercritical fluid carbon dioxide extraction of pesticides from foods (carrots, lettuce, peanut butter, hamburger, fortified butter fat and fortified potatoes) using pelletized diatomaceous earth (Hopper and King, 1991). Mixing of the sample with pelletized diatomaceous earth, which disperses the sample material and adsorbs water, allows samples ranging from 95% water to pure lipophilic oils to be efficiently extracted using supercritical carbon dioxide. A novel technique to screen for pesticide residues in meat products using SFE coupled with enzyme assays was reported by France and King (1991). Extracted analytes are collected in water and the resulting solution is tested for pesticide residues by enzyme assay. Two approaches were reported: alachlor-fortified lard and bovine liver were monitored by static SFE coupled with an enzyme immunoassay; and carbofuran-fortified frankfurters were monitored with an enzyme assay based on cholinesterase inhibition.

On-line supercritical fluid solid phase extraction and SFC for the trace analysis of pesticides in soybean oil and rendered fats was reported (Murugaverl and Voorhees, 1991). On-line extraction and clean-up are accomplished in one step by using supercritical carbon dioxide prior to chromatographic separation.

6.4.3 *Insecticides*

The feasibility of using SFE to recover insecticides from the environment was studied using chlorpyrifos methyl spiked on a wheat-kernel substrate (Campbell *et al.*, 1989). This insecticide was selected owing to its intermediate molecular weight, moderate polarity and volatility (allowing determination by GC). Multidimensional chromatography, consisting of coupled microcolumn liquid chromatography (LC) and GC with electron-capture detection (ECD), was used to effect clean-up of the extract and determine quantitatively the recoveries of the insecticide. Recovery experiments were done by using a 50 ng g^{-1} spike of the chlorpyrifos methyl on to the dried wheat kernel. After extraction (10 min at 300 atm and 50 °C) an aliquot (0.5 µl) was injected on to the LC column (fused silica packed with Spherisorb ODS) with an eluant of acetonitrile:water (85:15), and a flow rate of 6.0 µl min^{-1} (Figure 6.3). A 2.3 min cut from the LC column was

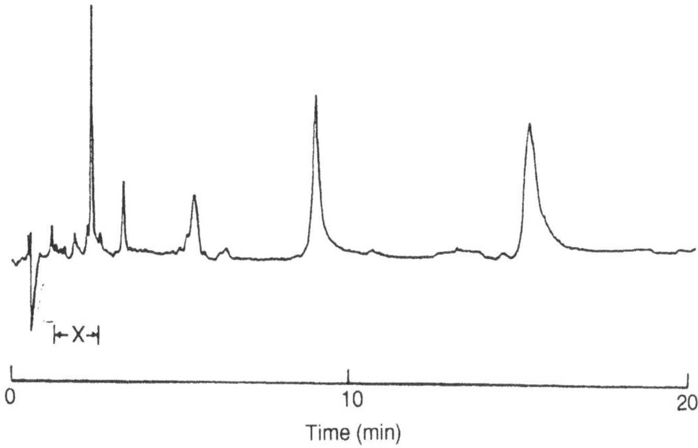

Figure 6.3 Reverse-phase liquid chromatography of an unspiked wheat-kernel extract. Column: 40 cm × 250 µm i.d. fused-silica packed with Spherisorb ODS; eluant: acetonitrile: water, 85:15; flow rate: 6 µl min^{-1}; detection: UV at 214 nm, 0.1 AUFS; injection: 0.5 µl; extraction conditions: CO_2, 50 °C and 300 atm; X = section transferred to the gas chromatograph. Reproduced with permission from Campbell *et al.* (1989).

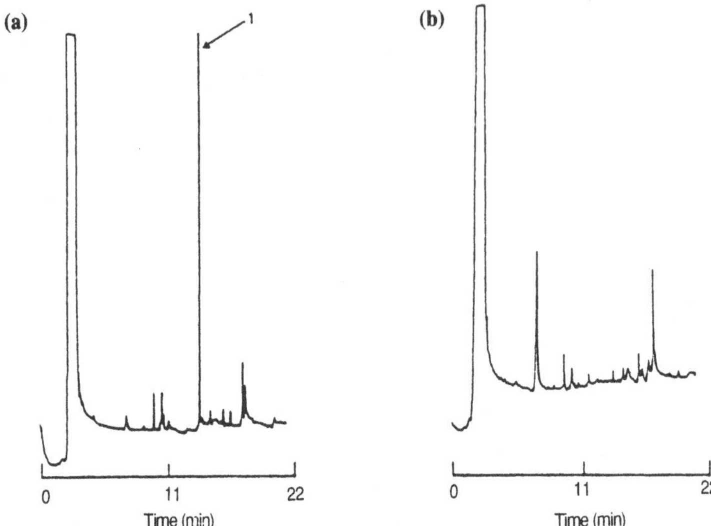

Figure 6.4 Gas chromatograms for: (a) a standard injection of 50 pg of chlorpyrifos methyl; and (b) an injection of an unspiked wheat-kernel extract with supercritical CO_2 at 50 °C and 300 atm. Temperature programme: 115–270 °C at 8 °C min^{-1}; uncoated inlet: 5 m × 0.25 mm i.d. undeactivated fused silica. Chromatograms represent the section transferred from the microcolumn liquid chromatogram (section X, Figure 6.3). Peak identification: (1) chlorpyrifos methyl. Reproduced with permission from Campbell *et al.* (1989).

introduced on to the GC column (undeactivated fused silica). Figure 6.4 shows the gas chromatograms for: (a) a standard injection of the insecticide, and (b) an unspiked wheat-kernel extract. The use of multidimensional chromatography allows clean-up (microcolumn LC) of the extract prior to GC–ECD. The resultant GC trace, Figure 6.4(b), is significantly less complex than that obtained when an extract was injected directly into the GC–ECD. However, when the multidimensional chromatography was applied to an extract of the wheat kernel spiked with chlorpyrifos methyl many more peaks were observed than with the unspiked wheat-kernel extract, Figure 6.4(b). This was reported to be due to the method of preparation used for the spiking experiment. In this situation the chlorpyrifos methyl spike was prepared in a few millilitres of solvent, methylene chloride. Although not reported, it is evident from subsequent work (Hills *et al.*, 1991) that the methylene chloride was acting as a modifier. The reported recoveries of the insecticide were improved from 65% to 98% when supercritical carbon dioxide modified with 2% methanol (supplied from a cylinder) was employed.

6.4.4 Herbicides

Two herbicides, linuron and diuron, have been studied by using directly coupled SFE–SFC (Wheeler and McNally, 1989). Initially, off-line SFE was optimized with respect to extraction-phase density at constant temperature, temperature at constant density, static equilibration time, and extraction-phase additives (primarily methanol and ethanol). Although the paper was concerned with the extraction of two herbicides from soil, some preliminary results were obtained for extraction from wheat grain. The workers concluded (for soil) that both herbicides could be extracted with >95% recovery provided a modifier was used. Recovery experiments were confirmed using radiolabelled ^{14}C compounds. However, each herbicide was modifier-specific: the highest extraction efficiencies for linuron were obtained with ethanol, while methanol proved to be the best for diuron. Earlier, the same group (McNally and Wheeler, 1988), had reported similar findings for the sulphonylurea herbicides, precursors and metabolites from complex matrices such as ground soybean, and several wheat matrices (grain, flour and straw).

The feasibility of using supercritical carbon dioxide in conjunction with a hydrolytic sample pretreatment for the extraction of 2,4-dichlorophenol residues from selected food crop tissues has been reported (Thomson and Chesney, 1992). 2,4-Dichlorophenol is a plant metabolite of the herbicides 2,4-dichlorophenoxyacetic acid and 2,4-dichlorophenoxybutanoic acid. 2,4-Dichlorophenol is also one of the eleven phenols designated as priority pollutants by the US Environmental Protection Agency (EPA). Current methodology for sample pretreatment is lengthy, involving liquid–liquid extraction, column clean-up and solvent evaporation after initial hydrolysis of the sample. Thus, as is the case for most, if not all, pollutants, a method is required that provides rapid and reliable extraction with the minimum number of sample-handling operations. This then has the potential for a method in which SFE can have a major role. Initial experiments using spiked samples of straw and seed with supercritical carbon dioxide (2300 psi at 40 °C for 1 h) showed good recoveries from straw (barley and triticale) whereas poorer recoveries were obtained from seed samples. However, field-treated barley straw samples gave extremely poor results (0.23 ppm compared to 1.84 ppm obtained by steam distillation). Further work was obviously required. This involved an investigation of potential sample pretreatment methods on a barley straw matrix. The pretreatment methods involved the addition of modifiers (ethanol and iso-octane), aqueous acid solution, aqueous base solution and/or heating of the sample. The best compromise pretreatment conditions were established as 17% H_3PO_4 at 100 °C for 4 h. This acid pretreatment procedure was believed to hydrolyse partially the plant tissue prior to extraction and to release the bound 2,4-dichlorophenol residue. However, for field-treated straw samples higher recoveries were obtained by steam distillation. In contrast, the converse was true for the field-treated seed samples, with SFE providing greater recoveries of 2,4-dichlorophenol. This was attributed to the high diffusivity of the supercritical fluid and its ability to diffuse into the seed interior, solvate the 2,4-dichlorophenol and hence aid migration to the seed surface. The total sample preparation time for either SFE or steam distillation was similar despite the difference in the extraction times of 45 min and 6 h, respectively. The similarity in total sample preparation time was principally attributed to the acid pretreatment time of 4 h required for SFE.

6.4.5 Mycotoxins

On-line SFE with chemical ionization mass spectrometry and collision-induced dissociation tandem mass spectrometry has been applied to the identification of trichothecene mycotoxins in wheat (Kalinoski *et al.*, 1986). Trichothecenes are a group of sesquiterpenoid mycotoxins produced by fungi. In this work three mycotoxins were studied, deoxynivalenol, diacetoxyscirpenol and T-2 toxin. The direct extraction of underivatized, unpurified grain samples without the need for modified supercritical carbon dioxide was demonstrated. Rapid detection and identification without exhaustive extraction was possible by the use of collision-induced dissociation tandem mass spectrometry.

6.5 Vitamins

Vitamin E or α-tocopherol is a fat-soluble and labile vitamin found principally in plant seed oils. This has led investigators to consider methods of enrichment from seeds. The beneficial properties of supercritical fluids allow extraction and chromatography to be used (Saito and Yamauchi, 1990*a,b*). Saito *et al.* (1989) described a method that incorporates three on-line processes: extraction with supercritical carbon dioxide, preconcentration, and fractionation by SFC. Tocopherols were extracted from wheatgerm powder (3 g) reported to contain 0.03% tocopherols and 10% oil by weight. The feasibility of extracting labile tocopherols in an oxygen-free environment at low temperature demonstrates one of the major advantages to be gained by using supercritical fluids. In addition, the incorporation of three on-line processes should allow this technique to replace the conventional normal-phase preparative HPLC approach.

Extraction by supercritical carbon dioxide of the lipophilic compound, mentadione (Vitamin K_3), from an animal feed (rat chow) was reported by Schneiderman *et al.* (1988). Mentadione (2-methyl-1,4-naphthoquinone) is added as a supplement to animal feed to prevent deficiency brought about by antibiotic or anticoagulant treatment. Current analytical isolation methods involve lengthy and time-consuming extraction processes which undoubtedly introduce analytical uncertainty. The method described shows the quantitative recovery of mentadione using supercritical carbon dioxide at 8000 psi and 60 °C for 20 min. The extracts were subsequently analysed without further sample clean-up using reverse-phase high performance liquid chromatography (HPLC) with reductive mode electrochemical detection at a silver electrode at −0.75 V versus calomel electrode. The minimum quantity extractable using this arrangement was 20 μg g^{-1} of mentadione in the feed.

6.6 Lipids

Capillary SFC provides a suitable technique for the separation of lipids from a variety of sources. Of those reported, the work of King (1990) has identified a variety of sample types to which SFC can be applied. Figure 6.5 shows a chromatogram of a fish oil capsule in which cholesterol and α-tocopherol (Vitamin E) are clearly separated. However, under the particular chromatographic conditions used the fish oil triglycerides are unresolved. In addition, SFC has been used to study the saponification of jojoba oil (Figure 6.6). This shows the versatility of SFC to study the conversion of agriculturally derived material and assess reactant–product mixtures.

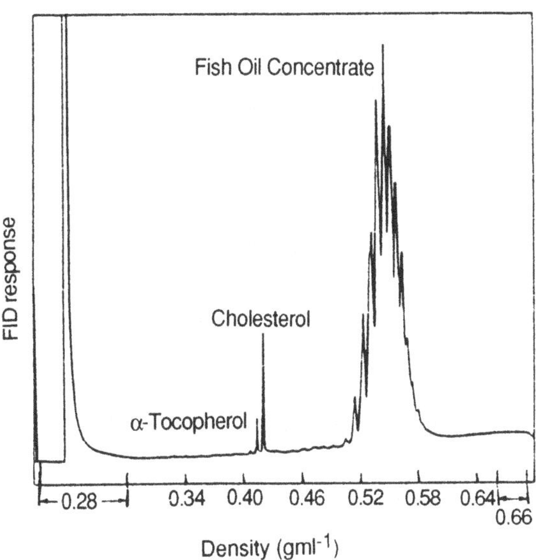

Figure 6.5 Determination of cholesterol and α-tocopherol (vitamin E) in a fish oil capsule by capillary supercritical fluid chromatography with a flame ionization detector. Column: 10 m × 50 μm i.d. SB-methyl (cross-linked methylpolysiloxane polymer); temperature: 120 °C; injection valve: Valco, cooled to −10 °C; injection volume: 200 nl; back-pressure restrictor: frit. Reproduced from King (1990) by permission of Preston Publications, a division of Preston Industries, Inc.

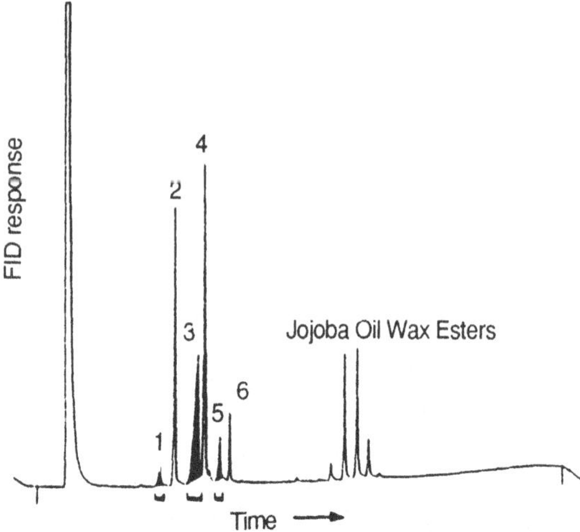

Figure 6.6 Capillary supercritical fluid chromatography separation of the fatty acids and alcohols from saponification of jojoba oil. Conditions and column were as in Figure 6.5. Jojoba oil has an average molecular weight of 606. Peak identification: (1) n-C_{18-19}; (2) n-$C_{20:1}$; (3) n-$C_{20:1}$; (4) n-$C_{22:1}$; (5) n-$C_{22:1}$; (6) n-$C_{24:1}$. Shaded product peaks are the acid moieties. The jojoba oil wax ester distribution is characterized by chain lengths from C_{38} to C_{46}. Reproduced from King (1990) by permission of Preston Publications, a division of Preston Industries, Inc.

An understanding of the processes that occur in seed during its ageing is a fundamental requirement in guaranteeing the quality of the seed provided and its potential for germination. The peroxidation of seed lipids, leading initially to hydroperoxide formation, has been proposed as a fundamental mechanism of deterioration during seed storage (Hannan and Hill, 1991). The processes involved were studied by using high-pressure liquefied gas (carbon dioxide) extraction because of its low temperature and inert atmosphere. Separation and detection were by capillary SFC with flame ionization detection (FID). The monitoring process involved a study of the lipid content of onion seeds exposed to two different storage conditions. The ageing process results in a chemical redistribution or fragmentation of the lipids. It has not been possible to observe this previously by using traditional fatty acid analysis, which involves hydrolysis of the fatty acids.

The ability to detect changes in the chemical composition of vegetable oils due to abuse or processing has been reported. Packed-microbore SFC with FID was reported by France *et al.* (1991*b*) for the monitoring of abused vegetable oils. The results, by the reported method for free fatty acid determination, compared favourably with a standard titration method. The use of SFC with on-line Fourier transform infrared (FTIR) detection to monitor the processing of vegetable oils (Calvey *et al.*, 1991) offers additional benefits due to the use of selective detection. Refined soybean oil and soybean oil that was partially hydrogenated were chosen for the study because of their dietary importance. The use of SFC with flow-cell FTIR allows the relative level of unsaturation and the extent of isomerization in partially hydrogenated soybean oil to be determined. Free fatty acids from the hydrolysis of soybean oil and the intact triacylglycerols of soybean oil were analysed. A major advantage of this analysis is the ability to monitor unsaturation and isomerization in a single chromatographic run. This is achieved by the selective monitoring facilitated by FTIR of the C–H deformation region (1000–900 cm^{-1}) in *trans*-$R_1HC=CHR_2$ groups, the C–H stretching region (3020–2800 cm^{-1}), and the carbonyl region (1800–1700 cm^{-1}).

6.6.1 *Hydroperoxide*

The use of SFE and SFC to study lipid peroxide levels in foods containing fats and oils has been reported (Sugiyama *et al.*, 1990). As discussed above, peroxide compounds are believed to accelerate the deterioration of the taste and appearance of food. Although standard methods of analysis are available, they often involve lengthy sample clean-up and extraction processes using Soxhlet extraction. Sugiyama *et al.* (1990) investigated the extraction and analysis of hydroperoxides from peanut oil samples using supercritical carbon dioxide. The combined use of SFE and SFC minimized sample handling. In this arrangement 40 mg of peanut powder (ground and sieved to a 60-mesh fineness) was spiked with the internal standard, stigmasterol, and extracted using supercritical carbon dioxide with 5% ethanol added at 140 bar and 40 °C for 7 min. This method compared favourably with the standard method of detection, potentiometry.

6.6.2 *Cholesterol*

A reliable, sensitive and accurate method of determining cholesterol is required because of its implication in coronary disease. A suitable study has been done using egg yolk as the sample (Ong *et al.*, 1990). This paper compared two methods of extracting cholesterol from egg yolk, SFE and Soxhlet extraction using *n*-hexane. Separation and detection was done using capillary SFC with an FID. With cholesteryl chloroacetate as an internal standard, quantitation was possible using SFC of the extracts. Both SFE and Soxhlet gave almost 100% recoveries from an 0.2 g egg-yolk sample spiked with 0.03 g internal standard. However, the extraction times between the two were considerably different, with SFE taking 1 h and Soxhlet 7 h. The additional advantage of SFE is the low temperature required for extraction. The use of a temperature such as 35 °C would allow other thermally labile compounds present in food samples, such as proteins, vitamins and nutrients, to be analysed.

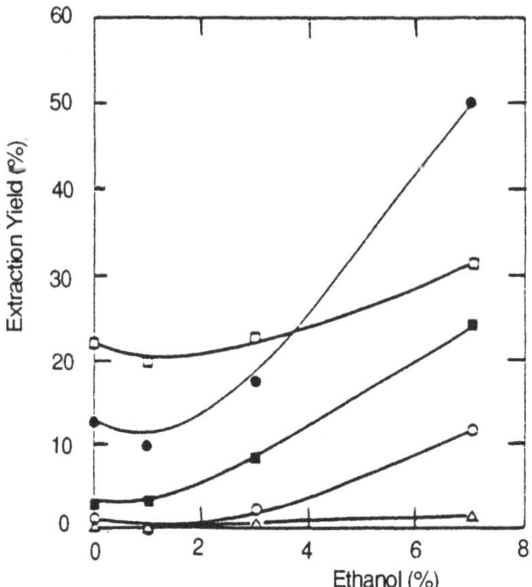

Figure 6.7 Extraction yields obtained during supercritical extraction of egg-yolk powder with reference to total solids (■), cholesterol (●), lecithin (□); cephalin (○) and proteins (△), as a function of ethanol concentration. Reproduced with permission from Rossi *et al.* (1990).

The extension to this work has been considered by Rossi *et al.* (1990). This group was able to use the selectivity of SFE to extract cholesterol and lecithin when using supercritical carbon dioxide without proteins being extracted, while the addition of an ethanol modifier (7%) favoured the slight extraction of proteins (Figure 6.7). In addition, total lipid extraction increased from 5% with supercritical carbon dioxide to 45% with the addition of 7% ethanol to the supercritical fluid.

6.6.3 Glycerides

An important consideration when optimizing the extraction of components from sample matrices is information relating to the solubility of the target analyte in the supercritical fluid. Although this may seem to be of fundamental importance only a limited amount of information is available in the literature. The solubility of an analyte in the supercritical fluid is not, however, subject to standard conditions, as is the case of a substance's solubility in, for example, water (i.e. 20 °C and 1 atm). One of the major advantages of a supercritical fluid is the variation in solvent strength with respect to temperature and pressure. Figure 6.8 shows this variability for triglycerides extracted from soybean meal (King, 1989). As can be noted, the general trend is for increasing temperature and pressure to lead to greater solubility of triglycerides in supercritical carbon dioxide. Similar

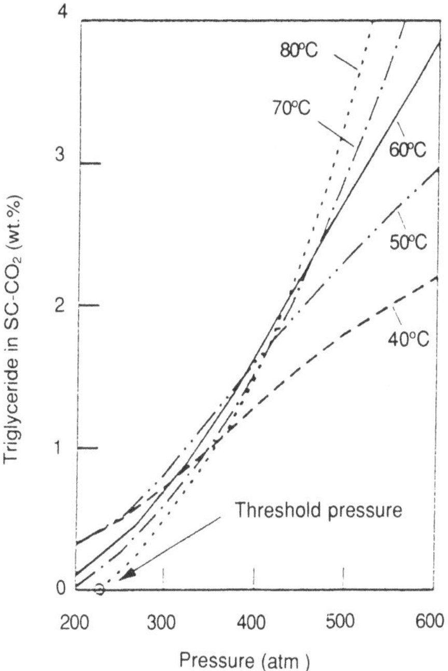

Figure 6.8 Solubility of triglycerides from soybean meal in supercritical carbon dioxide as a function of temperature and pressure. Reproduced from King (1989) by permission of Preston Publications, a division of Preston Industries, Inc.

work has investigated the solubility of some important cheese constituents, including triglycerides, in supercritical carbon dioxide (Gmur *et al.*, 1986). This knowledge is part of the fundamental basis for developing a model for extraction with supercritical fluids.

The use of supercritical fluids to extract triglycerides from natural products has been reported from a group at the University of New South Wales, Australia (Tilly *et al.*, 1990; Wells *et al.*, 1990). This group investigated the variables (temperature, 40–80 °C; pressure, 100–300 bar) that affect the extraction of triglycerides present in vegetable oils. The solubility of the triglycerides was found to be dependent upon the solvent density and the solute volatility. The problems of limited solubility (1–3 wt%) and high operating pressure (300–500 bar) with associated high capital and operating costs prevent the commercial use of supercritical carbon dioxide for triglyceride extraction. Therefore, an alternative supercritical fluid is required that can improve the extraction process but at lower cost. An initial report (Wells *et al.*, 1990) suggests that supercritical propane may be the solvent of choice for the extraction of triglycerides from natural products. Perhaps this optimism should be balanced by the major disadvantage of using propane, its flammability, with explosive limits in air of 2.37–9.5% by volume.

Quantitative approaches in capillary SFC are often avoided because of the inherent variation in injecting reproducible amounts (nanolitres) on to the capillary column. A preliminary study (White and Houck, 1985) investigated the choice of column for the analysis of mono-, di- and triglycerides using FID. The two different stationary phases investigated were DB-5 (95% dimethyl- (5%) diphenylpolysiloxane) and DB-225 (50% cyanopropylmethyl (50%) methylphenylpolysiloxane). Subsequently, Proot *et al.* (1986) reported the effect on peak resolution as a function of column temperature (90, 110, 150, 210 and 250 °C) for the analysis of triglycerides. Temperatures between 150 and 230 °C were recommended to avoid peak broadening and peak splitting that occurred above 230 °C. No evidence for thermal degradation of the triglycerides was reported at any temperature investigated.

A study involving the determination of mono- and diglycerides with and without derivatization has been reported (Lee *et al.*, 1991). Quantitation was achieved by injecting known concentrations of monomyristin, monopalmitin, monostearin, dimyristin, dipalmitin and distearin on to the column (SB-methyl-100) and recording the response factor (unit weight per peak area) of each analyte. The response factor was then used in the quantitation of the analytes in the test samples from their respective integrated peak areas. An internal standard (monomyristin) was used for the quantitation of monoglycerides in commercial emulsifiers.

6.7 Colours

One of the major advantages of SFE over other extraction techniques is the ability to control the solvent strength. This was ably demonstrated in a short paper in which the chromophore of a plant material was extracted at different carbon dioxide densities and monitored spectrophotometrically (Kane *et al.*, 1992). The extension of this work to food colours can easily be made.

Carotenoid pigments are responsible for the orange and yellow colours present in fruit and vegetables, e.g. carrot. Their labile nature makes them ideal compounds for extraction and chromatography by supercritical fluids. Supercritical carbon dioxide was used to extract carotene and lutein from leaf protein concentrates (Favati *et al.*, 1988). Extractions were done at pressures of 10–70 MPa and 40 °C. Over 90% of the carotene contained in the leaf protein concentrate was removed at pressures in excess of 30 MPa. However, 70% extraction of lutein from the sample required higher pressure (70 MPa). The reported method allows the possibility of selective extraction of natural colours, free of solvent residuals, which can then be used as food dyes. A recent study of food colours with spectrophotometric detection using near-critical carbon dioxide has been reported (Jay *et al.*,

1991). In this work a range of carotenoids and chlorophylls was examined. As expected, their solubility was temperature- and pressure-dependent. In addition, no chemical activity between supercritical carbon dioxide and carotene was observed.

In the more classical role of extraction followed by chromatography, carotenoids have been widely studied (Schmitz *et al.*, 1989; Aubert *et al.*, 1991; Lesellier *et al.*, 1991). In the first of these papers Schmitz *et al.* (1989) extracted carotenes from tomato and carrots and subsequently analysed the extracts by capillary SFC. Extraction was done by blending 200 g fresh weight of the vegetable in 200 ml of methanol for 5 min. The resultant purée was filtered. The remaining vegetable material was extracted with hexane (200 ml) for 30 min under an atmosphere of argon. Finally, rotary evaporation of the material produced a concentrated extract of 5 ml. Figure 6.9 shows the SFC separation (50 °C under isobaric conditions and a mobile-phase density of 0.70 g ml^{-1}) of α- and β-carotenes from carrot extract. Although not discussed, the logical progression of this work would be to see if extraction can be affected using SFE with carbon dioxide, maintaining an inert atmosphere at low temperature.

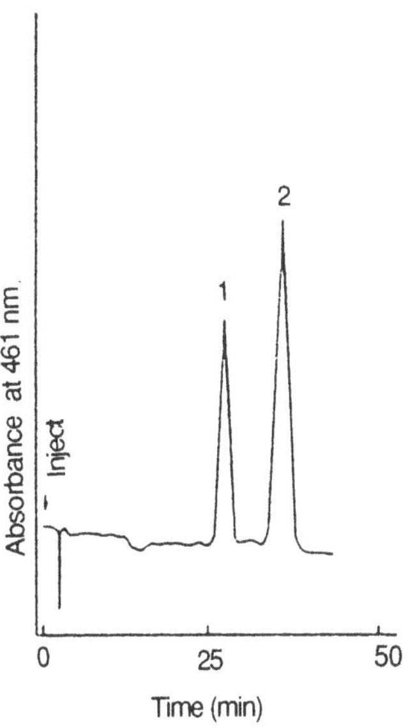

Figure 6.9 Capillary supercritical fluid chromatography of α- and β-carotene from carrot extract. Column: 10 m × 50 μm i.d. SB-phenyl (cross-linked stationary phase containing 50% phenyl- and 50% methylpolysiloxane); conditions: 50 °C under isobaric conditions and a mobile phase density of 0.70 g ml^{-1}; injection: Valco valve; injection volume: 200 nl internal loop in a time-split mode (0.25 s); detection: linear UV–VIS 204. Peak identification: (1) α-carotene; (2) β-carotene. Reproduced from Schmitz *et al.* (1989) by permission of Elsevier Science Publishers B.V.

A comprehensive study of *trans* and *cis* α- and β-carotenes by SFC has been reported (Aubert *et al.*, 1991; Lesellier *et al.*, 1991). In the first of these two papers (Aubert *et al.*, 1991) the effects of temperature, pressure and organic modifiers on the retention of carotenes were studied. The range of modifiers studied included methanol, acetonitrile, tetrahydrofuran, dichloromethane and trichlorotrifluoroethane, used as binary or ternary mixtures with carbon dioxide in the range 3–20% (v/v). This initial paper was followed (Lesellier *et al.*, 1991) by a detailed investigation of the affect that the type of octadecyl-bonded stationary phase (twenty-two were studied) has on the retention and selectivity of carotenes.

6.8 Sugars

The analysis of polar compounds, such as sugars, is potentially difficult in SFC. However, several approaches are possible involving the selection of the mobile phase and the column. An earlier approach involved derivatization (Chester and Innis, 1986) to facilitate elution of polar sugars, oligo- and polysaccharides containing up to 18 glucose units. Subsequently, the application of SFC with polar packed columns and light-scattering detection for the analysis of sugars has been reported (Herbreteau *et al.*, 1990). Cyano-, diol- and nitro-bonded silicas were used with modified carbon dioxide as the mobile phase. The use of packed columns with a methanol-modified mobile phase decreases the retention time of polar solutes. Use of this combination (packed column and modified carbon dioxide) renders the FID useless, so unless a chromophore is present, allowing use of spectrophotometry, detection of eluting species is a problem. This group has avoided this difficulty by using a light-scattering detector. The retention times of a range of mono- and disaccharides were studied by using methanol–carbon dioxide at a temperature of 40 °C. Figure 6.10 shows the separation of a range of sugars on two different columns, diol- and nitro-bonded silicas.

6.9 Waxes

The major interest in waxes in food science relates to the use of paper laminated with wax to preserve food. A typical example would be the waxed paper coating of some cheeses. In this situation it is possible for the waxes to migrate to the food surface and directly enter the food chain. Other uses of waxes include cosmetics, drum and vat liners, chewing gum, candles, soap and other minor applications. Capillary SFC and FID provides a suitable technique to profile and characterize waxes. In our laboratory, a series of wax profiles have been obtained including paraffin,

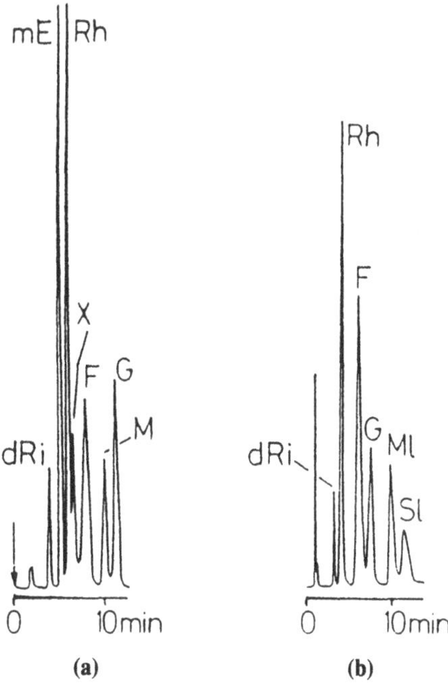

Figure 6.10 Separation of sugars using supercritical fluid chromatography packed columns. (a) Lichrospher diol column; eluant: carbon dioxide–methanol (84.5:15.5, w/w); flow rate: 1.77 ml min^{-1}; pressure: 3900 psi. (b) RSil NO_2 column; eluant: carbon dioxide–methanol (87.0:13.0, w/w); flow rate: 3.8 ml min^{-1}; pressure; 3500 psi. Peak identification: dRi, 2-deoxy-D-ribose; mE, meso-erythritol; Rh, rhamnose; X, xylose; F, fructose; M, mannose; G, glucose; Ml, mannitol; Sl, sorbitol. Reproduced from Herbreteau *et al.* (1990) by permission of Elsevier Science Publishers B.V.

beeswax, Japan, Candelilla, Carnauba, Ceresin, Mantan, microcrystalline and Ozekerite. Figures 6.11 and 6.12 show typical capillary SFC traces for selected waxes. Similar chromatograms have also appeared in the scientific literature, most notably the work of Hawthorne and Miller (1987).

6.10 Flavours and aromas

Flavours and aromas have provided a major application area for supercritical fluids in food science, but it is not necessarily the most significant. This is because the majority of papers in the scientific and manufacturers' literature have used this area of investigation to demonstrate the simplicity, selectivity and effectiveness of, in particular, SFE. This should not, however, distract us from the importance of extracting flavour and aromas, but should influence our choice of sample-preparation technique (an issue not lost on the marketing departments of instrument manufacturers).

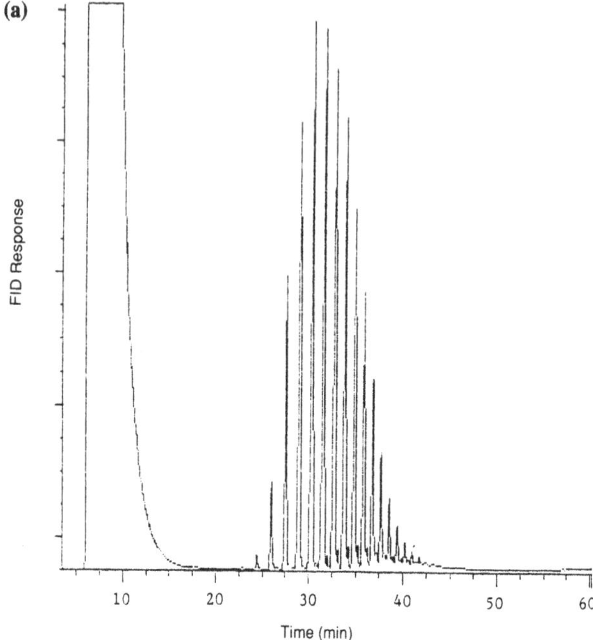

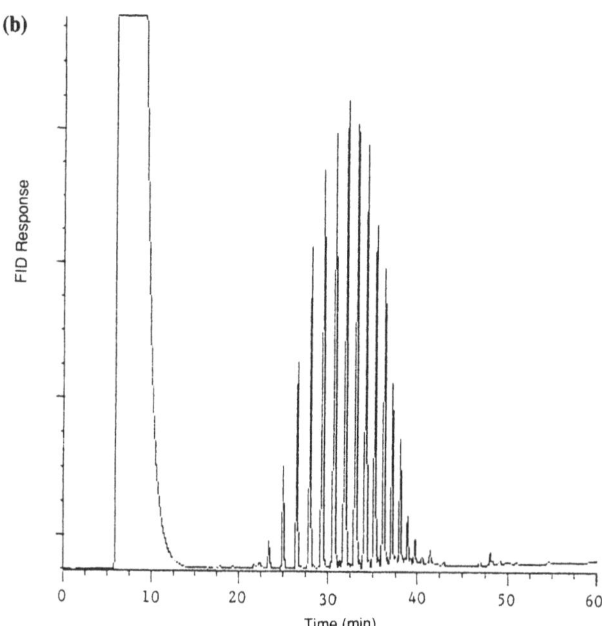

Figure 6.11 Capillary supercritical fluid chromatography with flame ionization detection of: (a) paraffin wax; and (b) white Ceresin wax. Column: 10 m × 50 μm i.d. SB-phenyl-30 (Dionex); eluant: carbon dioxide; programme, 0.25–0.80 g ml^{-1} at 0.01 MPa min^{-1}; temperature: 90 °C; injector: Valco valve; injection volume: 200 nl in a time-split (40 ms); back-pressure restrictor: frit.

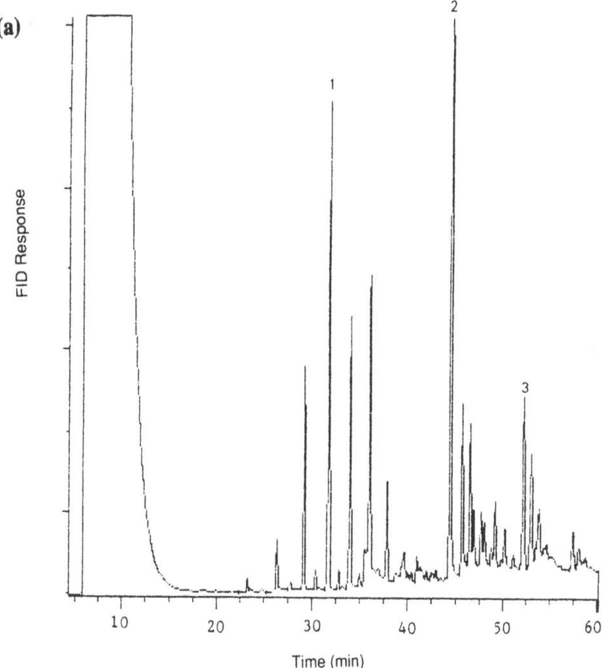

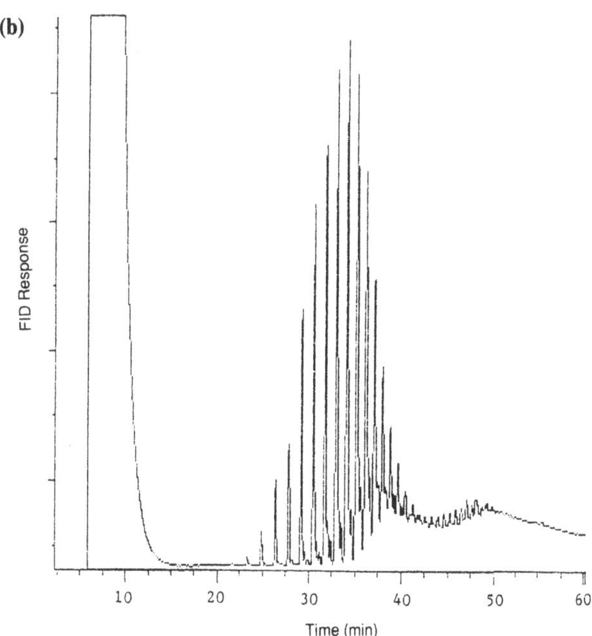

Figure 6.12 Capillary supercritical fluid chromatography with flame ionization detector of (a) beeswax and (b) Ozokerite wax. For conditions, see Figure 6.11. Peak identification: (1) alkanes (molecular weight 350–470); (2) esters (molecular weight 590–750); (3) diesters (molecular weight 850–1000).

6.10.1 *Caffeine*

Of the common demonstrations that show the applicability of analytical SFE, this is surely the most common and often quoted example. It is also the best known commercial application of SFE (caffeine from coffee) (Jenkins *et al.*, 1991). Several publications describe the conditions and experimental arrangement necessary to extract caffeine from coffee or tea (Williams, 1981; Sugiyama *et al.*, 1985; Skelton *et al.*, 1986; Anton *et al.*, 1988; Engelhardt and Gross, 1988; Sandra *et al.*, 1990; Hills *et al.*, 1991; Jenkins *et al.*, 1991). Since we do not wish to avoid this topic, an HPLC trace from one of the earlier analytical papers on the topic is shown in Figure 6.13 (Sugiyama *et al.*, 1985). Perhaps of more importance is the overview given in the paper of Jenkins *et al.* (1991), which demonstrates the versatility and applicability of FTIR spectroscopy to detect not only caffeine in the coffee extract but other components as well. While the use of FTIR is a developing area in the field of supercritical fluids, in terms of

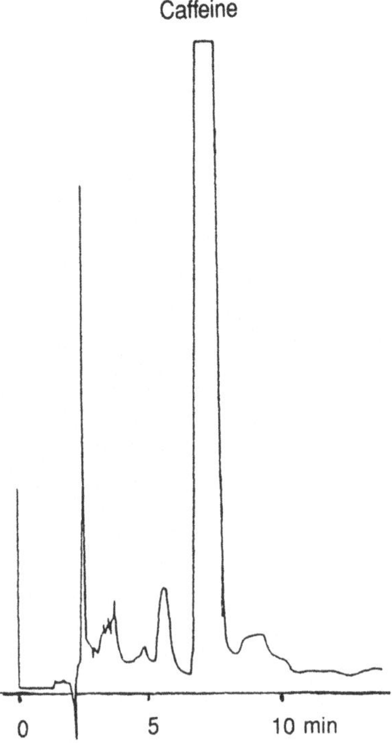

Figure 6.13 High performance liquid chromatogram of coffee extract by supercritical fluid extraction. SFE conditions: pressure, 200 bar; temperature, 48 °C; added water, 20%, time, 60 min; HPLC conditions: column, Jasco Fine Pak SIL C_{18}; eluant, methanol–water (55:45); flow rate, 1.2 ml min^{-1}; UV monitored at 272 nm and 0.64 AUFS. Reproduced from Sugiyama *et al.* (1985) by permission of Elsevier Science Publishers B.V.

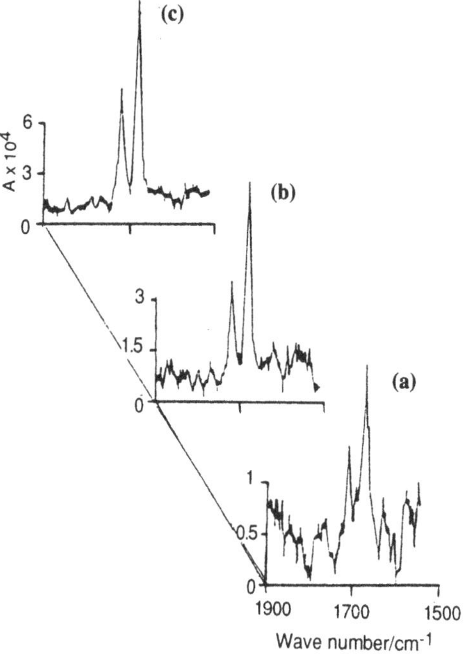

Figure 6.14 Fourier transform infrared spectra of the carbonyl stretching region of caffeine (1500–1900 cm^{-1}) to determine the injected minimum detection quantity in capillary supercritical fluid chromtography–Fourier transform infrared. Column: 10 m × 50 μm i.d., 5% phenyl-substituted methylpolysiloxane; temperature, 100 °C; eluant, carbon dioxide; programme, linear pressure ramp from 9.12 to 40.52 MPa at 1.01 MPa min^{-1}; injection, 100 nl direct, solution in CH_2Cl_2; detection, FTIR flow cell (0.98 μl) at 25 °C, quantity injected: (a) 95 pg, (b) 290 pg, and (c) 580 pg. Reproduced with permission from Jenkins *et al.* (1991).

the choice of supercritical fluid, instrumentation configuration and spectra obtained, the potential of the technique is enormous. Figure 6.14 shows the IR stretch of carbonyl recorded in supercritical carbon dioxide using capillary SFC with on-line FTIR detection. It is clearly observed that <100 pg can be quantitatively determined (Figure 6.14(a)).

Even more diverse is the identification of a caffeine extract by FTIR and Fourier transform–nuclear magnetic resonance (FT–NMR) spectrometry (Elisabeth *et al.*, 1991). Roasted coffee beans yielded approximately 25% of extract after SFE (40 °C, 200 kg cm^{-2}); after dilution with methanol the extract was subjected to preparative SFC. Solvent evaporation preceded identification of the caffeine extract by FTIR and FT–NMR.

6.10.2 Plant products

The extraction and/or chromatographic separation by supercritical fluids in conjunction with either more traditional sample preparation schemes, hyphenated techniques, spectrometry and chromatography merely outline the versatility of supercritical fluids. To this end a whole range of plant products has been analysed including: Chinese herbal medicines, frankincense, myrrh and *Evodia rutaecarpa* (Ma *et al.*, 1991); rosemary, thyme, cinnamon, spruce needle, cedar wood and orange peel (Hawthorne *et al.*, 1988*a*); eucalyptus leaves, lime peel, lemon peel and basil (Hawthorne *et al.*, 1989); rosemary, sage and orange peel (Hawthorne *et al.*, 1988*b*); brewing hops (Hawthorne *et al.*, 1990); cotton seed kernel (King, 1990); paprika (Skelton *et al.*, 1986); caraway seed (Engelhardt and Gross, 1988); Mexican spices, *Origanum vulgare* and *Pimpinella anisum* (Ondarza and Sanchez, 1990); hops, nutmegs and chillies (Williams, 1981); capsicum (Knowles *et al.*, 1988); and corn-germ flour (Christianson *et al.*, 1984).

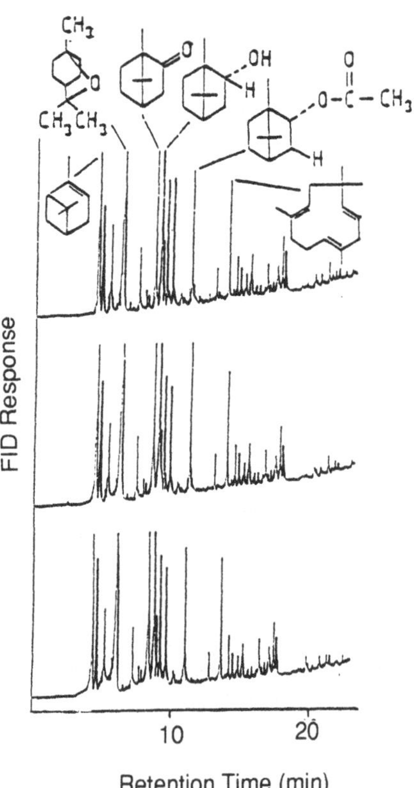

Figure 6.15 Triplicate analyses of rosemary by supercritical fluid extraction directly coupled with gas chromatography and flame ionization detection. Coupling of SFE was achieved by inserting the restrictor (15 cm × 15–30 μm i.d.) into the on-column injector. Extracted species (10 min at 45 °C and 300 atm using carbon dioxide) were cryogenically focused (−30 °C) on to a GC column: 30 m × 0.32 mm i.d. DB-5 with a 1 μm film thickness; temperature ramp of 8 °C min^{-1} from 70 °C to 320 °C. Reproduced with permission from Hawthorne *et al.* (1988*b*).

One of the figures to merit concern when evaluating a new technique is the reproducibility. Hawthorne *et al.* (1988*b*) demonstrated the reproducibility of SFE by extracting three 0.5 mg samples of rosemary for 10 min using supercritical carbon dioxide. The GC–FID traces of the replicate analyses are shown in Figure 6.15. Second extractions on all samples yielded no additional peaks, indicating that 10 min was sufficient to remove the extractable organics. Other work, however (see section 6.1), has indicated that the addition of a modifier (methanol) or derivatizing agent would probably extract further components.

6.10.3 Essential oils

Several papers have focused on the extraction and separation of essential oils using supercritical fluids (Williams, 1981; Christianson *et al.*, 1982, 1984; Friedrich and List, 1982; Friedrich *et al.*, 1982; Friedrich and Pryde, 1984; List *et al.*, 1984*a,b*; Snyder *et al.*, 1984; List and Friedrich, 1985, 1989; Pubols *et al.*, 1985; Hawthorne *et al.*, 1988*a*, 1989, 1990; Knowles *et al.*, 1988; Sugiyama and Saito, 1988; Anderson *et al.*, 1989; Kassim and Hameed, 1990; Favati *et al.*, 1991). This is because of the need to improve the existing methods of analysis in terms of speed of analysis and quality of the final result. Essential oils are widely used in the flavour and fragrance industries. The composition of the essential oil is obviously one of the main factors that determine the quality of the final product and hence its retail cost. The variability in the essential oil composition due to climatic and source variations makes quality-control checks a prerequisite. Variability can also arise due to the selection of the analytical technique, method of extraction and deterioration of the oil upon storage (Sugiyama and Saito, 1988). These factors have led to investigators assessing the potential of supercritical fluids as effective agents in extraction and separation techniques.

The selectivity of SFE to extract effectively different components of a sample at different densities has been applied to a cold-pressed grapefruit oil (Anderson *et al.*, 1989). By careful control of the extraction density of the supercritical carbon dioxide they were able to extract selectively the oxygen-containing components from the hydrocarbon compounds. It was reported that it is the oxygen-containing substituents of the essential oil that provide the desired aroma, and that the unstable hydrocarbon compounds (sesquiterpenes) can lead to premature deterioration and loss of quality of the oil. However, 90% of the oil is made up of the hydrocarbon fraction. Figure 6.16 demonstrates the effectiveness of SFE to exclude the higher-molecular-weight sesquiterpenes from the oxygen fraction at low density (0.1767 g ml^{-1}).

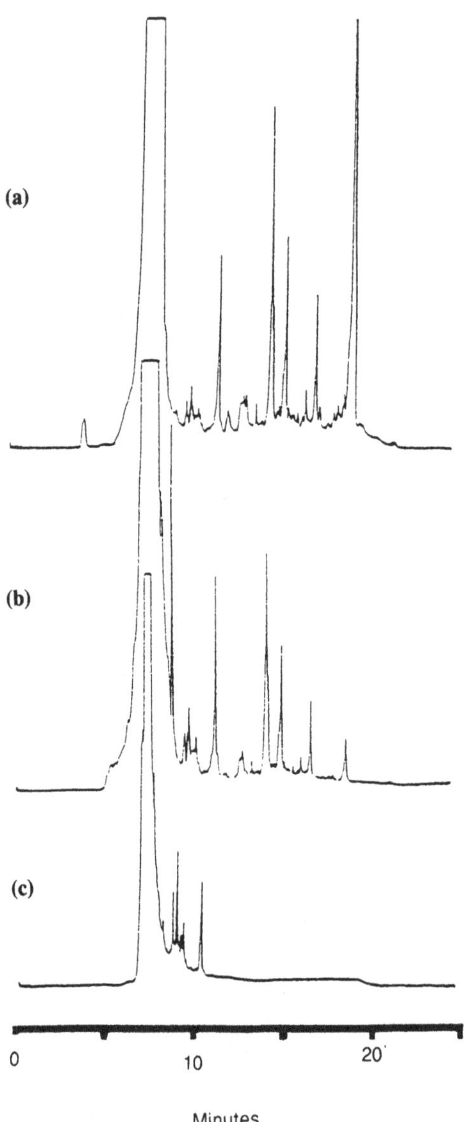

Figure 6.16 Supercritical fluid extraction–supercritical fluid chromatography with flame ionization detection of cold-pressed grapefruit oil at extraction densities (g ml^{-1}) of (a) 0.8579; (b) 0.3560; (c) 0.1767. Extraction: 12 min at 70 °C with carbon dioxide; cryotrapping at −65 °C; column; 10 m × 100 μm SB-phenyl-5; eluant: carbon dioxide; programme, linear density from 0.1005 to 0.6100 g ml^{-1} at 0.01 g ml^{-1} min^{-1} and 150 °C. Reproduced from Anderson *et al.* (1989) by permission of Preston Publications, a division of Preston Industries, Inc.

Acknowledgements

I acknowledge the help of Jerry W. King of the United States Department of Agriculture, Peoria, Illinois, in providing valuable references from his group relating to the use of the supercritical fluids in food science, and of Mr E. Ludkin, University of Northumbria at Newcastle, for the chromatograms shown in Figures 6.11 and 6.12. Finally, I acknowledge financial support from the following organisations: University of Northumbria at Newcastle, ICI plc, Glaxo Manufacturing Services, and Analytical and Environmental Services, Northumbrian Water.

References

Anderson, M.R., Swanson, J.T., Porter, N.L. and Richter, B.E., (1989), Supercritical fluid extraction as a sample introduction method for chromatography. *J. Chromatogr. Sci.*, **27**, 371–377.

Anton, K., Menes, R. and Widmer, H.M., (1988), Direct coupling of carbon dioxide fluid extraction with capillary supercritical fluid chromatography. *Chromatographia*, **26**, 221–223.

Aubert, M.C., Lee, C.R., Krstulovic, A.M., Lesellier, E., Pechard, M.R. and Tchapla, A., (1991), Separation of trans/cis- α- and β-carotenes of supercritical fluid chromatography. I. Effects of temperature, pressure and organic modifiers on the retention of carotenes. *J. Chromatogr.*, **557**, 47–58.

Calvey, E.M., McDonald, R.E., Page, S.W., Mossoba, M.M. and Taylor, L.T., (1991), Evaluation of SFC/FTIR for examination of hydrogenated soybean oil. *J. Agric. Food Chem.*, **39**, 542–548.

Campbell, R.M., Meunier, D.M. and Cortes, H.J., (1989), Supercritical fluid extraction of chlorpyrifos methyl from wheat at part per billion levels. *J. Microcol. Sep.*, **1**, 302–308.

Chester, T.L. and Innis, D.P., (1986), Separation of oligo- and polysaccharides by capillary supercritical fluid chromatography. *J. High Resolut. Chromatogr.*, **9**, 209–212.

Christianson, D.D., Friedrich, J.P., Bagley, E.B. and Inglett, G.E., (1982), Maize germ flours for food purposes by supercritical carbon dioxide extraction. In *Maize: Recent Progress in Chemistry and Technology*, Inglett, G.E., (Ed.). Academic Press, San Diego, CA, USA, pp. 231–239.

Christianson, D.D., Friedrich, J.P., List, G.R., Warner, K., Bagley, E.B., Stringfellow, A.C. and Inglett, G.E., (1984), Supercritical fluid extraction of dry-milled corn germ with carbon dioxide. *J. Food Sci.*, **49**, 229–232.

Cole, L.A., Dorsey, J.G. and Chester, T.L., (1991), Investigation of derivatizing agents for polar solutes in supercritical fluid chromatography, *Analyst*, **116**, 1287–1291.

Elisabeth, P., Yoshioka, M., Yamauchi, Y. and Saito, M., (1991), Infrared and nuclear magnetic resonance spectrometry of caffeine in roasted coffee beans after separation by preparative supercritical fluid chromatography. *Anal. Sci.*, **7**, 427–431.

Engelhardt, H. and Gross, A., (1988), On-line extraction and separation by supercritical fluid chromatography with packed columns. *J. High Resolut. Chromatogr.*, **11**, 38–42.

Favati, F., King, J.W., Friedrich, J.P. and Eskins, K., (1988), Supercritical carbon dioxide extraction of carotene and lutein from leaf protein concentrates. *J. Food Sci.*, **53**, 1532–1536.

Favati, F., King, J.W. and Mazzanti, M., (1991), Supercritical carbon dioxide extraction of evening primrose oil. *J. Amer. Oil Chem. Soc.*, **68**, 422–427.

France, J.E. and King, J.W., (1991), Supercritical fluid extraction/enzyme assay: A novel technique to screen for pesticide residues in meat products. *J. Assoc. Off. Anal. Chem.*, **74**, 1013–1016.

France, J.E., King, J.W. and Snyder, J.M., (1991*a*), Supercritical fluid-based cleanup technique for the separation of organochlorine pesticides from fats. *J. Agric. Food Chem.*, **39**, 1871–1874.

France, J.E., Snyder, J.M. and King, J.W., (1991*b*), Packed-microbore supercritical fluid chromatography with flame ionization detection of abused vegetable oils. *J. Chromatogr.*, **540**, 271–278.

Friedrich, J.P. and List, G.R., (1982), Characterization of soybean oil extracted by supercritical carbon dioxide and hexane. *J. Agric. Food. Chem.*, Jan–Feb., 192–193.

Friedrich, J.P. and Pryde, E.H., (1984), Supercritical carbon dioxide extraction of lipid-bearing materials and characterization of the products. *J. Amer. Oil Chem. Soc.*, **61**, 223–228.

Friedrich, J.P., List, G.R. and Heakin, A.J., (1982), Petroleum-free extraction of oil from soybeans with supercritical carbon dioxide. *J. Amer. Oil. Chem. Soc.*, **59**, 288–292.

Gmur, W., Bosset, J.O. and Plattner, E., (1986), Solubility of some important cheese constituents in supercritical carbon dioxide. *Lebensm. Wiss. Technol.*, **19**, 419–425.

Hannan, R.M. and Hill, H.H., Jr., (1991), Analysis of lipids in ageing seed using capillary supercritical fluid chromatography. *J. Chromatogr.*, **547**, 393–401.

Hawthorne, S.B. and Miller, D.J., (1987), Analysis of commercial waxes using capillary supercritical fluid chromatography–mass spectrometry. *J. Chromatogr.*, **388**, 397–409.

Hawthorne, S.B., Krieger, M.S. and Miller, D.J., (1988*a*), Analysis of flavor and fragrance compounds using supercritical fluid extraction coupled with gas chromatography. *Anal. Chem.*, **60**, 472–477.

Hawthorne, S.B., Miller, D.J. and Krieger, M.S., (1988*b*), Rapid extraction and analysis of organic compounds from solid samples using coupled supercritical fluid extraction/gas chromatography. *Fresenius Z. Anal. Chem.*, **330**, 211–215.

Hawthorne, S.B., Miller, D.J. and Krieger, M.S., (1989), Coupled SFE–GC: A rapid and simple technique for extracting, identifying, and quantitating organic analytes from solids and sorbent resins. *J. Chromatogr. Sci.*, **27**, 347–354.

Hawthorne, S.B., Miller, D.J. and Langenfeld, J.J., (1990), Quantitative analysis using directly coupled supercritical fluid extraction–capillary gas chromatography (SFE–GC) with a conventional split/splitless injection port. *J. Chromatogr. Sci.*, **28**, 2–8.

Hawthorne, S.B., Miller, D.J., Nivens, D.E. and White, D.C., (1992), Supercritical fluid extraction of polar analytes using *in situ* chemical derivatization. *Anal. Chem.*, **64**, 405–412.

Hedrick, J.L. and Taylor, L.T., (1989), Quantitative supercritical fluid extraction and chromatography of a phosphonate from aqueous media. *Anal. Chem.*, **61**, 1986–1988.

Hedrick, J.L. and Taylor, L.T., (1990), Supercritical fluid extraction strategies of aqueous based matrices. *J. High Resolut. Chromatogr.*, **13**, 312–316.

Herbreteau, B., Lafosse, M., Morin-Allory, L. and Dreux, M., (1990), Analysis of sugars by supercritical fluid chromatography using polar packed columns and light-scattering detection. *J. Chromatogr.*, **505**, 299–305.

Hills, J.W., Hill, H.H., Jr. and Maeda, T., (1991), Simultaneous supercritical fluid derivatization and extraction. *Anal. Chem.*, **63**, 2152–2155.

Hopper, M.L. and King, J.W., (1991), Enhanced supercritical fluid extraction of pesticides from foods using pelletized diatomaceous earth. *J. Assoc. Off. Anal. Chem.*, **74**, 661–666.

Jakobsson, M., Sivik, B., Bergqvist, P.A., Strandberg, B., Hjelt, M. and Rappe, C., (1991), Extraction of dioxins from cod liver oil by supercritical carbon dioxide. *J. Supercrit. Fluids*, **4**, 118–123.

Jay, A.J., Steytler, D.C. and Knights, M., (1991), Spectrophotometric studies of food colors in near-critical carbon dioxide. *J. Supercrit. Fluids*, **4**, 131–141.

Jenkins, T.J., Kaplan, M., Simmonds, M.R., Davidson, G., Healy, M.A. and Poliakoff, M., (1991), Novel methods of instrumental analysis using supercritical fluid based systems. An overview. *Analyst*, **116**, 1305–1311.

Kalinoski, H.T., Udseth, H.R., Wright, B.W. and Smith, R.D., (1986), Supercritical fluid extraction and direct fluid injection mass spectrometry for the determination of trichothecene mycotoxins in wheat samples. *Anal. Chem.*, **58**, 2421–2425.

Kane, M., Dean, J.R., Hitchen, S.M., Tranter, R.L. and Dowle, C.J., (1991). Analysis of surfactants using supercritical fluid chromatography and extraction. *European Symposium on Analytical Supercritical Fluid Chromatography and Extraction*, Wiesbaden, Germany, 4–5 December, p. 20.

Kane, M., Dean, J.R., Hitchen, S.M., Dowle, C.J. and Tranter, R.L., (1992), Supercritical fluid extraction as a sample preparation technique for chromatography and spectroscopy. *Anal. Proc.*, **29**, 31–33.

Kassim, D.M. and Hameed, M.S., (1990), Direct extraction-separation of essential oils from citrus peels by supercritical carbon dioxide. *Sep. Sci. Technol.*, **24**, 1427–1435.

King, J.W., (1989), Fundamentals and applications of supercritical fluid extraction in chromatographic science. *J. Chromatogr. Sci.*, **27**, 355–364.

King, J.W., (1990), Applications of capillary supercritical fluid chromatography–supercritical fluid extraction to natural products. *J. Chromatogr. Sci.*, **28**, 9–14.

Knowles, D.E., Richter, B.E., Wygant, M.B., Nixon, L. and Anderson, M.R., (1988), Supercritical fluid chromatography: A new technique for AOAC. *J. Assoc. Off. Anal. Chem.*, **71**, 451–457.

Lee, T.W., Bobik, E. and Malone, W., (1991), Quantitative determination of mono- and diglycerides with and without derivatization by capillary supercritical fluid chromatography. *J. Assoc. Off. Anal. Chem.*, **74**, 533–537.

Lesellier, E., Tchapla, A., Pechard, M.R., Lee, C.R. and Krstulovic, A.M., (1991), Separation of *trans/cis* α- and β-carotenes by supercritical fluid chromatography. II. Effect of the type of octadecyl-bonded stationary phase on retention and selectivity of carotenes. *J. Chromatogr.*, **557**, 59–67.

List, G.R. and Friedrich, J.P., (1985), Processing characteristics and oxidative stability of soybean oil extracted with supercritical carbon dioxide at 50 °C and 8000 psi. *J. Amer. Oil Chem. Soc.*, **62**, 82–84.

List, G.R. and Friedrich, J.P., (1989), Oxidative stability of seed oils extracted with supercritical carbon dioxide. *J. Amer. Oil Chem. Soc.*, **66**, 98–100.

List, G.R., Friedrich, J.P. and Christianson, D.D., (1984*a*), Properties and processing of corn oils obtained by extraction with supercritical carbon dioxide. *J. Amer. Oil. Chem. Soc.*, **61**, 1849–1851.

List, G.R., Friedrich, J.P. and Pominski, J., (1984*b*), Characterization and processing of cottonseed oil obtained by extraction with supercritical carbon dioxide. *J. Amer. Oil Chem. Soc.*, **61**, 1847–1849.

Liu, H., Cooper, L.M., Raynie, D.E., Pinkston, J.D. and Wehmeyer, K.R., (1992), Combined supercritical fluid extraction/solid-phase extraction with octadecylsilane cartridges as a sample preparation technique for the ultratrace analysis of a drug metabolite in plasma. *Anal. Chem.*, **64**, 802–806.

Ma, X., Yu, X., Zheng, Z. and Mao, J., (1991), Analytical supercritical fluid of Chinese herbal medicines. *Chromatographia*, **32**, 40–44.

McNally, M.E. and Wheeler, J.R., (1988), Supercritical fluid extraction coupled with supercritical fluid chromatography for the separation of sulphonylurea herbicides and their metabolites from complex matrices. *J. Chromatogr.*, **435**, 63–71.

Moulder, R., Bartle, K.D. and Clifford, A.A., (1991), Coupled microcolumn size exclusion liquid chromatography–capillary supercritical fluid chromatography. *Analyst*, **116**, 1293–1298.

Murugaverl, B. and Voorhees, K.J., (1991), On-line supercritical fluid extraction/chromatography system for trace analysis of pesticides in soybean oil and rendered fats. *J. Microcol. Sep.*, **3**, 11–16.

Ondarza, M. and Sanchez, A., (1990), Steam distillation and supercritical fluid extraction of some Mexican spices. *Chromatographia*, **30**, 16–18.

Ong, C.P., Lee, H.K. and Li, S.F.Y., (1990), Supercritical fluid extraction and chromatography of cholesterol in food samples. *J. Chromatogr.*, **515**, 509–513.

Proot, M., Sandra, P. and Geeraert, E., (1986), Resolution of triglycerides in capillary SFC as a function of column temperature. *J. High Resolut. Chromatogr.*, **9**, 189–192.

Pubols, M.H., McFarland, D.C., Eldridge, A.C. and Friedrich, J.P., (1985), Feed efficiency and pancreatic enzymes of chickens fed soybean meal extracted with supercritical carbon dioxide. *Nutrition Reports International*, **31**, 1191–1200.

Ramsey, E.D., Perkins, J.R., Games, D.E. and Startin, J.R., (1989), Analysis of drug residues in tissue by combined supercritical fluid extraction–supercritical fluid chromatography–mass spectrometry–mass spectrometry. *J. Chromatogr.*, **464**, 353–364.

Rossi, M., Specicato, E. and Schiraldi, A., (1990), Improvement of supercritical carbon dioxide extraction of egg lipids by means of ethanolic entrainer. *Ital. J. Food. Sci.*, **4**, 249–255.

Saito, M. and Yamauchi, Y., (1990*a*), *Supercritical Fluid Extraction and Supercritical Fluid Chromatography for Preparative Separation*. Jasco Report Publisher, Tokyo, Japan, Chap. 7, pp. 125–142.

Saito, M. and Yamauchi, Y., (1990*b*), *Supercritical Fluid Extraction and Supercritical Fluid Chromatography for Preparative Separation*. Jasco Report Publisher, Tokyo, Japan, Chap. 9, pp. 155–177.

Saito, M., Yamauchi, Y., Inomata, K. and Kottkamp, W., (1989), Enrichment of tocopherols in wheat germ by directly coupled supercritical fluid extraction with semipreparative supercritical fluid chromatography. *J. Chromatogr. Sci.*, **27**, 79–85.

Sandra, P., David, F. and Stottmeister, E., (1990), Recovery studies by off-line SFE. *J. High Resolut. Chromatogr.*, **13**, 284–286.

Schmitz, H.H., Artz, W.E., Poor, C.L., Dietz, J.M. and Erdman, J.W., Jr., (1989), High-performance liquid chromatography and capillary supercritical fluid chromatography separation of vegetable carotenoids and carotenoid isomers. *J. Chromatogr.*, **479**, 261–268.

Schneiderman, M.A., Sharma, A.K. and Locke, D.C., (1988), Determination of mentadione in an animal feed using supercritical fluid extraction and HPLC with electrochemical detector. *J. Chromatogr. Sci.*, **26**, 458–462.

Skelton, R.J., Johnson, C.C. and Taylor, L.T., (1986), Sampling considerations in supercritical fluid chromatography. *Chromatographia*, **21**, 3–8.

Snyder, J.M., Friedrich, J.P. and Christianson, D.D., (1984), Effect of moisture and particle size on the extractability of oils from seeds with supercritical carbon dioxide. *J. Amer. Oil Chem. Soc.*, **61**, 1851–1856.

Sugiyama, K. and Saito, M., (1988), Simple microscale supercritical fluid extraction system and its application to gas chromatography–mass spectrometry of lemon peel. *J. Chromatogr.*, **442**, 121–131.

Sugiyama, K., Saito, M., Hondo, T. and Senda, M., (1985), New double-stage separation analysis method. Directly coupled laboratory scale supercritical fluid extraction–supercritical fluid chromatography, monitored with a multiwavelength ultraviolet detector. *J. Chromatogr.*, **332**, 107–116.

Sugiyama, K., Shiokawa, T. and Moriya, T., (1990), Application of supercritical fluid chromatography and supercritical fluid extraction to the measurement of hydroperoxides in foods. *J. Chromatogr.*, **515**, 555–562.

Thomson, C.A. and Chesney, D.J., (1991), Supercritical carbon dioxide extraction of 2,4-dichlorophenol from food crop tissues. *Anal. Chem.*, **64**, 848–853.

Tilly, K.D., Chaplin, R.P. and Foster, N.R., (1990), Supercritical fluid extraction of the triglycerides present in vegetable oils. *Sep. Sci. Technol.*, **25**, 357–367.

Wells, P.A., Foster, N.R., Liong, K.K. and Chaplin, R.P., (1990), Supercritical fluid extraction of triglycerides. *Sep. Sci. Technol.*, **25**, 139–154.

Wheeler, J.R. and McNally, M.E., (1989), Supercritical fluid extraction and chromatography of representative agricultural products with capillary and microbore columns. *J. Chromatogr. Sci.*, **27**, 534–539.

White, C.M. and Houck, R.K., (1985), Analysis of mono-, di- and triglycerides by capillary supercritical fluid chromatography. *J. High Resolut. Chromatogr.*, **8**, 293–296.

Williams, D.F., (1981), Extraction with supercritical gases. *Chem. Eng. Sci.*, **36**, 1769–1788.

7 Supercritical fluid extraction in environmental analysis

V. JANDA, K. BARTLE and A.A. CLIFFORD

7.1 Introduction

Supercritical fluid extraction (SFE) is a novel sample preparation method which promises to have a profound influence on environmental analytical chemistry, since it affords rapid, selective and convenient sample clean-up (McHugh and Krukonis, 1986; Hawthorne, 1990; Lee and Markides, 1990; Westwood, 1992). The main advantage of a supercritical fluid over an extracting liquid is that its properties, namely density, solvating power, viscosity and solute diffusivity, can all be controlled by varying the applied pressure. This leads to a greater selectivity, rapid mass-transfer and higher flow-rates as compared with liquids. Further, the separation of solvent from solute is simply achieved by decompression, since the solvent is usually gaseous at ambient temperature. SFE also has considerable advantages over liquid extraction in terms of sample size, cost and volume of solvent and analysis time. Extraction can be performed either off-line or on-line; couplings of SFE to gas, liquid and supercritical fluid chromatographs have all been demonstrated.

7.2 Practice

Table 7.1 lists the critical parameters of a number of substances for possible use in environmental analysis (Lee and Markides, 1990). Carbon dioxide has so far been the most widely used because of its convenient critical properties, non-toxicity, cheapness and non-flammable character. It is usually classified as a non-polar solvent, but its large quadrupole moment leads to some affinity with polar solutes and many large polar organic molecules are soluble in it.

For the extraction of more polar molecules, polar modifiers such as those listed in Table 7.2 are usually added to the CO_2 (Lee and Markides, 1990). The modifier phase diagram must be considered to ensure that there will be only one phase under the conditions of the extraction. Thus for methanol–CO_2 at 50 °C, there is only one phase above 95 bar whatever the composition, but below this pressure two phases are possible. A comprehensive

Table 7.1 Critical parameters of selected substances useful as supercritical fluids

	T_c (°C)	P_c (atm)	ρ_c (10^3 kg m^{-3})
CO_2	31.3	72.9	0,47
N_2O	36.5	72.5	0.45
SF_6	45.5	37.1	0.74
NH_3	132.5	112.5	0.24
H_2O	374	227	0.34
n-C_4H_{10}	152	37.5	0.23
n-C_5H_{12}	197	33.3	0.23
Xe	16.6	58.4	1.10
CCl_2F_2	112	40.7	0.56
CHF_3	25.9	46.9	0.52

compilation of phase data for CO_2 mixed with numerous modifiers has been published (Page *et al.*, 1992).

Experimental SFE is conceptually simple (Hawthorne, 1990): a pump is used to supply a known pressure of the extraction fluid to an extraction cell held at a temperature above the critical temperature of the fluid, which then flows through the sample and out through a pressure restrictor to a collecting device at atmospheric pressure (Figure 7.1). The pump is either a syringe or a reciprocating type with a cooled head. Modifier, if required, is introduced either by means of a separate pump via a mixing device, or from a premixed cylinder; it should be remembered, however, that if a

Table 7.2 Modifiers for carbon dioxide in SFE

Modifier	T_c (°C)	P_c (atm)	Molecular mass	Dielectric constant at 20 °C	Polarity index
Methanol	239.4	79.9	32.04	32.70	5.1
Ethanol	243.0	63.0	46.07	24.3	4.3
1-Propanol	263.5	51.0	60.10	20.33	4.0
2-Propanol	235.1	47.0	60.10	18.3	3.9
1-Hexanol	336.8	40.0	102.18	13.3	3.5
2-Methoxyethanol	302	52.2	76.10	16.93	5.5
Tetrahydrofuran	267.0	51.2	72.11	7.58	4.0
1,4-Dioxane	314	51.4	88.11	2.25	4.8
Acetonitrile	275	47.7	41.05	37.5	5.8
Dichloromethane	237	60.0	84.93	8.93	3.1
Chloroform	263.2	54.2	119.38	4.81	4.1
Propylene carbonate	352.0		102.09	69.0	6.1
N,*N*-Dimethylacetamide	384		87.12	37.78	6.5
Dimethyl sulphoxide	465.0		78.13	46.68	7.2
Formic acid	307		46.02	58.5	
Water	374.1	217.6	18.01	80.1	10.2
Carbon disulphide	279	78.0	76.13	2.64	

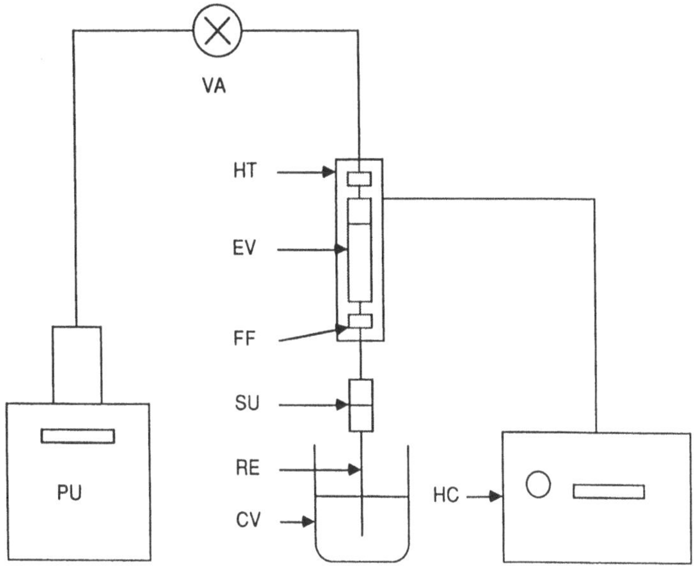

Figure 7.1 Off-line SFE system: PU, CO_2 pump; VA, on/off valve; HT, thermostatted heating tube; EV, extraction vessel; FF, finger-tight connectors; SU, swagelock union; RE, restrictor; CV, collection vial; HC, heater controller.

cylinder is used, the modifier composition changes slightly as the contents are consumed.

The solid sample to be extracted is held between frits in an extraction cell usually fabricated from stainless steel and available from a number of suppliers. The extraction-cell dimensions may affect the rate of extraction (Furton and Rein, 1991), perhaps because of turbulence effects; diffusers and ultrasonic irradiation have also been employed. Cell orientation and fluid flow direction are important if the cell is not full, but less important if it is. Liquid or wet samples may be mixed (King, 1990) with an adsorbent such as pelletised Celite (Hydromatrix) or a drying agent (e.g. magnesium sulphate). Organic water pollutants may also be adsorbed on to a solid adsorbent (either solid-phase extraction cartridges or filter discs) from which they are removed for chromatographic analysis by SFE (Wright *et al.*, 1987*b*). Direct SFE of aqueous solutions has also been demonstrated (Hedrick and Taylor, 1989).

The restrictor maintains the pressure within the cell. It may be, most simply, a length of fused silica tubing with an internal diameter between 20 and 50 μm or a crimped stainless-steel tube. More elaborate devices include back-pressure regulators and micrometer valves. The extracted material may also be collected in a vial containing solvent or by direct cooling. The restrictor is usually heated to prevent blockages when extracting materials containing water, which freezes as the supercritical fluid evaporates. This arrangement also prevents deposition of extracted material in the restrictor during extraction of sediments containing elemental sulphur; locating a copper scavenger column between cell and restrictor is also recommended (Pyle and Setty, 1991). Solid traps containing glass beads, silica gel or a liquid chromatographic stationary phase have also been used to collect analytes. Alternatively, SFE equipment is directly coupled to an analytical instrument, and such systems are discussed in section 7.4.

7.3 Factors affecting extraction from environmental samples

Three interrelated factors influence recovery during SFE as is shown in the so-called SFE triangle (Figure 7.2) (Westwood, 1992). For successful extraction, the solute must first be sufficiently soluble in the supercritical fluid. This is particularly important at the start when extraction is occurring rapidly. The onset of extraction in a graph of percentage recovery in a given time against fluid pressure or density (Figure 7.3(a)) is referred to as the threshold pressure. Control of solubility via applied pressure may allow stepwise extraction; for example (Raynor *et al.*, 1988), the extraction of two- and three-ring polycyclic aromatic hydrocarbons (PAHs) from a coal-derived solid occurred at a CO_2 pressure of 100 bar, whereas five-ring PAHs required a pressure of 200 bar (Figure 7.4). These observations have been correlated with calculated solubilities (Shilstone *et al.*, 1990). It is thus important to know the conditions under which the analyte is sufficiently soluble. In fact, solubility of a substance in a supercritical fluid is the sum of two factors: (i) the volatility of the substance; and (ii) the solvating effect of the supercritical fluid, which is a function of fluid density (Lee and Markides, 1990). It is noteworthy that a decrease in solubility may occur as a consequence of repulsive forces 'squeezing' the solute out of solution. A number of compilations of solubility data for supercritical fluids have appeared (e.g. Bartle *et al.*, 1991), from which the threshold pressure can be determined; such data are generally obtained by gravimetric measurements, although a more rapid chromatographic procedure has

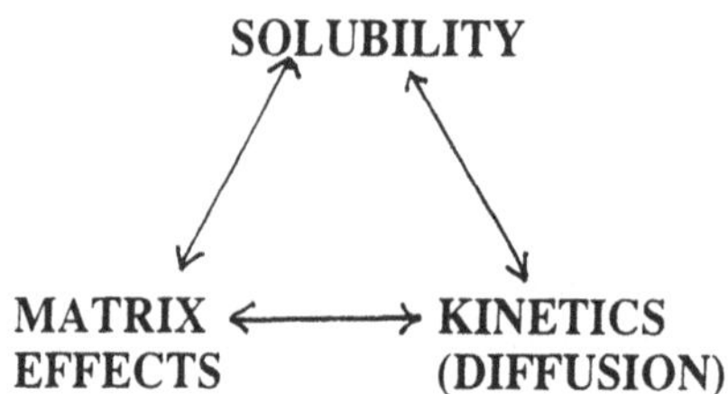

Figure 7.2 Supercritical fluid extraction triangle.

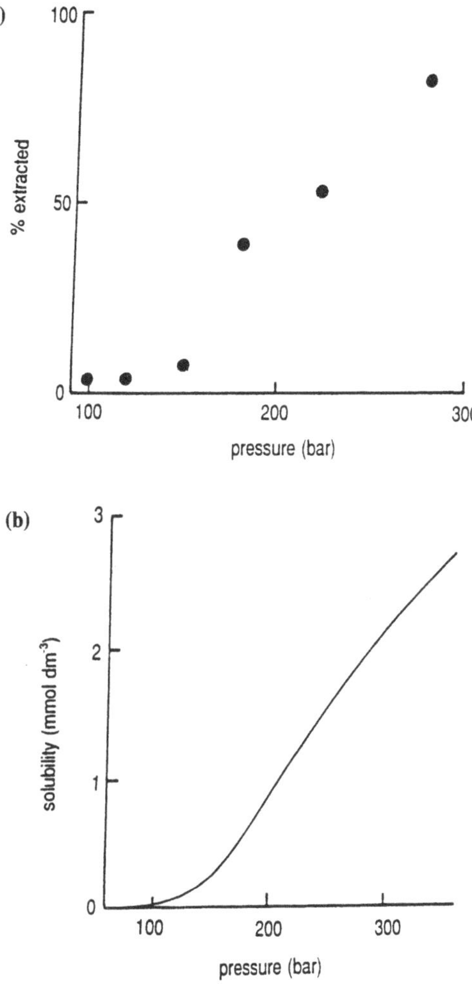

Figure 7.3 (a) Percentage recovery of atrazine from soil by SFE with carbon dioxide at different pressures after 15 min at 80 °C and constant flow rate; (b) calculated solubility at the same temperature (Ashraf *et al.*, 1992).

been used with some success (Bartle *et al.*, 1990*b*). Alternatively, supercritical fluid solubilities may be predicted either by use of an equation of state such as the Peng–Robinson equation (Bartle *et al.*, 1989), or by means of various empirical correlations so as to extend existing data. The solubility data for atrazine (Figure 7.3(b)) thus obtained (Ashraf *et al.*, 1992) correlate well with experimental extraction behaviour (Figure 7.3(a)): predicted solubilities begin to rise at 100 bar in agreement with the experimentally observed threshold solubilities. Janda *et al.* (1989*a*) observed that simazine was much less efficiently extracted from sediment than was atrazine. Calculated solubility curves for these compounds

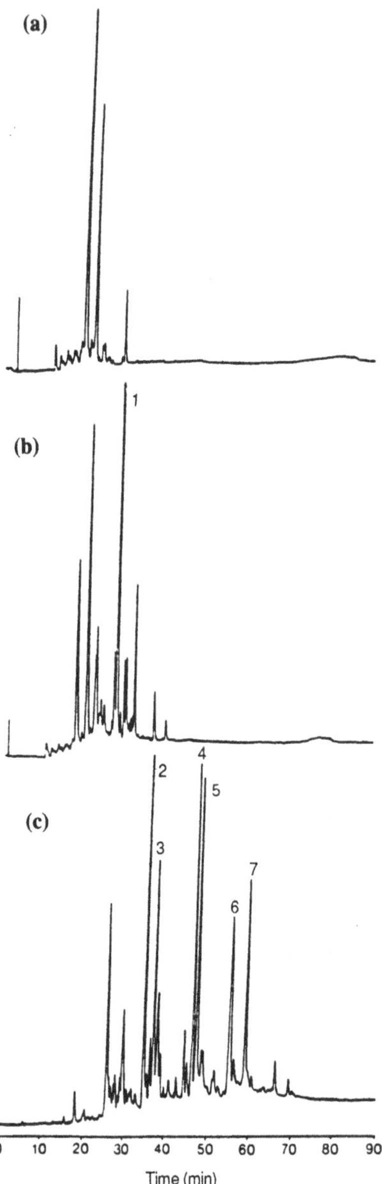

Figure 7.4 Supercritical fluid chromatography of coal-tar pitch extracted at: (a) 70 atm; (b) 100 atm; and (c) 200 atm. Peak identification: (1) phenanthrene; (2) fluoroanthene; (3) pyrene; (4) benz[*a*]anthracene; (5) chrysene; (6) benzofluoroanthene; (7) benzopyrene (Raynor *et al.*, 1988).

are in quantitative agreement with these results; simazine is predicted (Ashraf *et al.*, 1992) to be much less soluble than atrazine in supercritical CO_2.

SFE usually exhibits (Westwood, 1992) the time dependence shown in Figure 7.5. If the concentration of analyte in a continuous flow of fluid is well below the solubility limit, the rate-determining process is diffusion out of the matrix. An effective diffusion coefficient (D) and a particular matrix geometry are assumed, along with no solubility limitation; the solutions of the appropriate differential equations are obtained by the same methods as those applied to heat conduction. For a sphere, the solution is therefore described as the 'hot-ball' model (Bartle *et al.*, 1991). If the mass of solute, initially m_0, is m after time t, the ration m/m_0 is given by:

$$\frac{m}{m_0} = \frac{6}{\pi^2} \sum_{n=1}^{\infty} \frac{1}{n^2} \exp\left(-n^2 \pi^2 \frac{Dt}{r^2}\right) \tag{7.1}$$

where n is an integer. Making the substitution $\pi^2 \frac{D}{r^2} = a$,

$$\frac{m}{m_0} = \frac{6}{\pi^2}\left[\exp(-at) + \frac{1}{4}\exp(-4at) + \frac{1}{9}\exp(-9at) + \ldots\right] \tag{7.2}$$

representing a sum of exponential decays and reproducing (Figure 7.6(a)) observed behaviour. At long times, the later (faster decaying) terms decrease in importance in comparison with the first term; a graph of ln (m/m_0) versus t becomes linear (Figure 7.6(a)). Although the initial steeper fall appears as a small feature, it represents the extraction of the

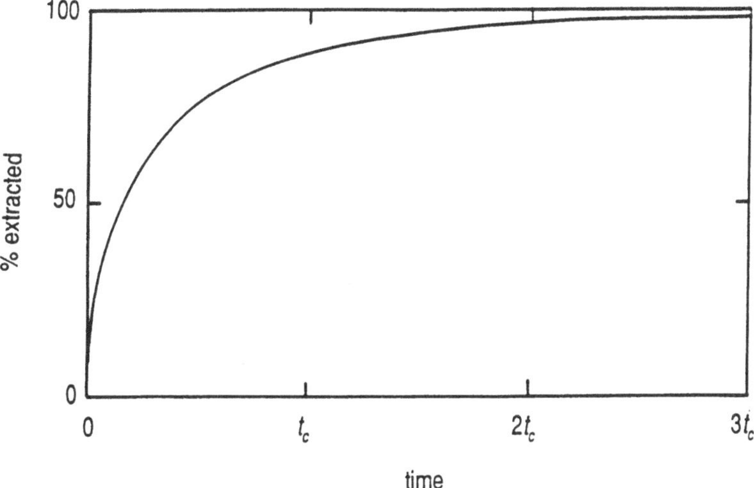

Figure 7.5 Theoretical curve of percentage extraction versus time.

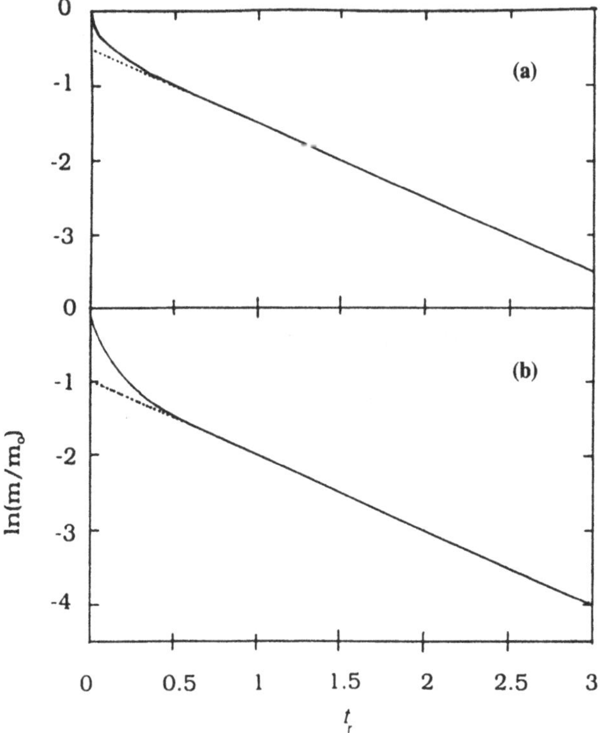

Figure 7.6 Graph of ln (m/m_0) versus scaled time for the hot-ball model: (a) basic model; (b) effect of irregular particle size.

majority of the material. Equation (7.1) shows a squared dependence on *r*, rationalising the well known fact that speed of extraction is increased by crushing or grinding solids, or coating liquids on a finely divided substrate. Real samples will contain particles of irregular shape and the curve has a large intercept compared with the value of −0.5 for spheres (Figure 7.6(b)).

Figure 7.7 shows experimental results (Bartle *et al.*, 1990*a*) for the extracton of phenanthrene from railway bed soil at 50 °C using CO_2 at 405 bar. The curve of ln (m/m_0) versus *t* has the form of the 'hot-ball' model although the intercept is close to −2, indicating irregular shapes for the soil particles. Similar curves have been found for the SFE of numerous analytes from environmental matrices, e.g. PAHs with molar masses from 128 to 252 (Janda *et al.*, 1989*a*) and atrazine (Ashraf *et al.*, 1992) from soils, and alkylbenzenesulphonates from organic digester sludge (Bartle *et al.*, 1990*a*).

The exponential behaviour of the extraction after the initial period means that extrapolation may be used to obtain quantitative analytical

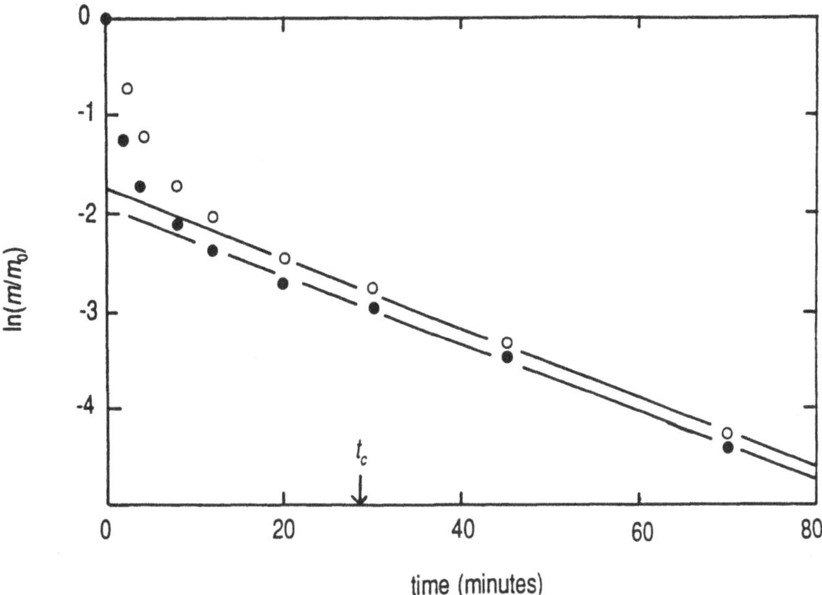

Figure 7.7 Supercritical fluid extraction of phenanthrene from railway soil. Extraction with carbon dioxide at 50 °C, pressures of 180 atm (○) and 400 atm (●), and constant flow rate (Bartle *et al.*, 1990*a*).

information without exhaustive extraction. If m_1 is the mass extracted in the initial non-exponential period and m_2 and m_3 are the masses extracted in two subsequent equal time periods, then the total mass, m_0, is given by:

$$m_0 = m_1 + \frac{m_2^2}{m_2 - m_3} \qquad (7.3)$$

This method has been tested by Liu *et al.* (1992) by coupled supercritical fluid extraction–gas chromatography (SFE–GC) of volatile organic compounds at sub-ppm levels in soil. Table 7.3 contains results of repeated runs during which extraction was carried out for approximately 20 min but not to completion. An overall RSD of 4.9% was observed. Table 7.4 compares the effect of different time intervals on the values of m_0 for the extraction of three compounds from soil. The calculated m_0 values were not affected as long as the first time period covers the entire non-linear region.

Solubility affects the kinetics of SFE since, as already stated, the 'hot-ball' (Bartle *et al.*, 1990*a*) model assumes no solubility limitations. If the concentration of solute in the fluid is finite (assumed to be proportional to the concentration in the matrix at the surface), and the partition coefficient is proportional to the solubility (S) of the solute in the fluid(s), a new

Table 7.3 Use of the extrapolation procedure during SFE of volatile organics from soil

	Measurement (arbitrary units)			
	m_1	m_2	m_3	m_0
	726	248	116	1190
	888	208	68	1200
	771	176	64	1050
	762	208	84	1110
	720	212	92	1100
	753	188	80	1080
	861	200	76	1180
	768	192	80	1100
	813	224	88	1180
Mean	785	206	83	1130
Standard deviation	58	21	15	55
RSD%	7.4	10	18	4.9

version of equation (7.1) can be written (Bartle *et al.*, 1992):

$$\frac{m}{m_0} = 6 \sum_{n=1}^{\infty} \frac{\left(\frac{hr}{ar}\right)}{\{hr(hr-1)+a_n^2\}\exp(-a_n^2Dhr^2)} \tag{7.4}$$

where a_n are the roots of the equation

$$a\cot(a) = 1 - hr$$

and $h=KSF/AD$, K is a constant for a particular matrix, A is its surface area and F is the volume flow rate of fluid. Plots of ln (m/m_0) against t now form a family of curves (Figure 7.8) for which the gradients of the linear

Table 7.4 Effect of time interval used in computing m_0

Time intervals used (s)			Measured value of m_0 (arbitrary units)		
Δt_1	Δt_2	Δt_3	Toluene	*p*-Xylene	Benzylamine
0–200	200–400	400–600	116	196	251
0–400	400–600	600–800	116	196	279
0–400	400–800	800–1200	116	197	266
0–300	300–800	800–1300	116	197	262
0–300	300–700	700–1100	116	196	263
0–500	500–1000	1000–1500	116	198	262
Mean m_0			116	197	266
Standard deviation (SD)			—	0.82	7.2
RSD%			—	0.42	2.7

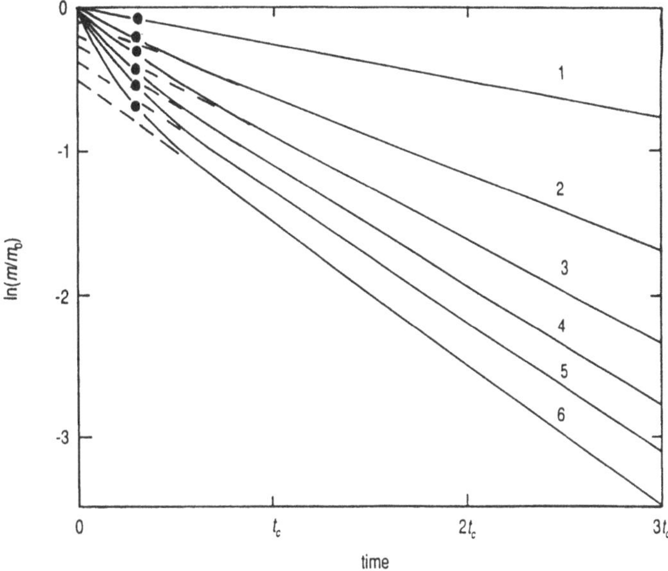

Figure 7.8 Solubility—limited hot-ball model plots (Bartle *et al.*, 1992).

portions become increasingly steep as solubility increases, with a limiting value for infinite solubility—the previously mentioned 'hot-ball' model.

These considerations are in keeping with effects observed during the SFE at 50 °C and constant flow rate of phenanthrene from soil (Figure 7.7). At 180 atm (182 bar) the ln (m/m_0) versus t curve falls less steeply initially than that at 400 atm (403 bar), and the linear portion is displaced upwards. The lower rate of extraction at 180 atm can be explained by the difference in solubility at the two pressures. At 50 °C, the saturated mole fraction of phenanthrene in CO_2 is 0.0015 at 180 atm but 0.003 at 400 atm. Although the two curves appear similar, the amount extracted after 4 min differs by over 10%.

7.4 On-line coupling of SFE to chromatographic analysis

Since the extraction solvent is easily removed, the analyte may be trapped for subsequent separation and trace analysis by directly coupled gas chromatography (GC), supercritical fluid chromatography (SFC) and high performance liquid chromatography (HPLC). In direct on-line coupling, the solvent peak is eliminated and the analysis of compounds which may elute with the solvent becomes possible. On-line coupling also reduces sample handling and the possibility of sample loss and contamination. The process is attractive if only limited amounts of sample are available since all of the analyte can be transferred.

A number of application areas of SFE–GC and SFE–SFC have been described (Hawthorne, 1990; Lee and Markides, 1990). If analytes are both thermally stable and volatile, then GC is the preferred separation technique; thus fuels, polychlorinated biphenyls and PAHs in environmental samples can all be analysed by on-line SFE–GC. When the sample contains thermally unstable or reactive compounds, SFE–SFC is recommended; SFE–SFC is particularly attractive (Westwood, 1992) since the extracting fluid may be the same as the mobile phase. SFE–SFC can be applied to the analysis of soils and sediments for compounds such as thermally unstable pesticides.

7.4.1 SFE–GC

On-line SFE–GC has the following steps: extraction, depressurisation and venting of the supercritical fluid, collection and focusing on to a GC column and subsequent GC analysis (Figure 7.9). Two broad methods of collecting and focusing have been reported (Hawthorne, 1990; Westwood, 1992). In the past, the extract was collected in an external device, e.g. by depressurising into a cold trap located before the GC; the trap was heated in a flow of carrier gas to transfer the extract to the column. Alternatively, the GC system may be used as the trap by depressurising into a retention gap before the column or by depositing the extract directly inside the column.

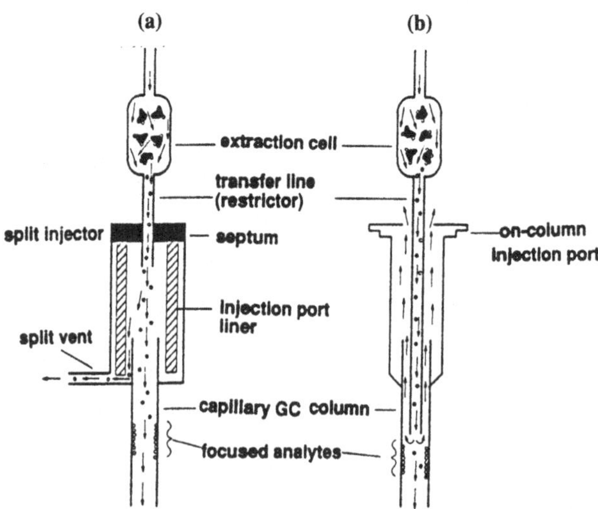

Figure 7.9 On-line coupling of supercritical fluid extraction to gas chromatography: (a) split SFE–GC; (b) on-column SFE–GC.

The cold-trapping procedure prevents any deleterious effects of the supercritical fluid on the GC column or detector but is limited by the efficiency of collection of volatiles in the extract. Use of a suitable adsorbent, e.g. Tenax, with thermal desorption may be used. SFE–GC with column collection is more efficient since the stationary phase helps to concentrate the extract in a narrow band; both on-column (Figure 7.9(b)) and split-injection (Figure 7.9(a)) versions have been reported for quantitative analysis. The former permits parts per billion detection limits for milligram amounts of sample extracted. Split injection is preferred if the sample is wet, since freezing and plugging of the restrictor outlet or GC column may occur with on-column injection.

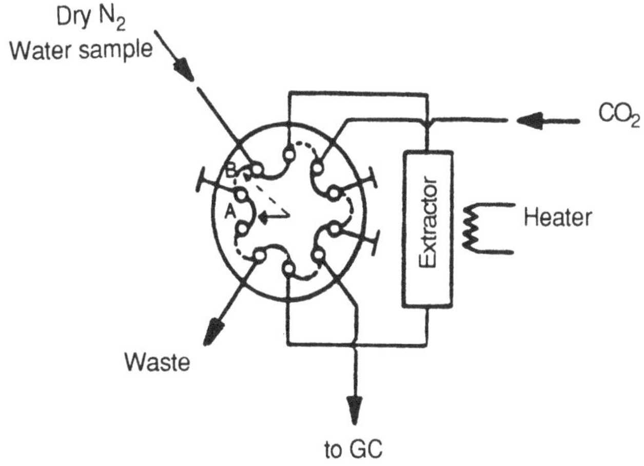

Figure 7.10 Apparatus for automated supercritical fluid extraction–gas chromatographic analysis of water (Pawlyszin and Alexandrou, 1989).

The coupling of SFE with GC provides a rapid and convenient method for the analysis of air and water matrices. Trace components in air may be concentrated on to adsorbents such as Tenax, polyurethane foam, charcoal or silica/alumina, from which they may be extracted by SFE and transferred quantitatively to the GC via the SFE–GC interface. Similarly, pollutants in water may be concentrated on polyurethane foam or solid-phase extraction cartridges and Empore discs and then analysed by SFE–GC.

An automated version of the latter equipment has been described (Pawliszyn and Alexandrou, 1989) (Figure 7.10), in which water was passed through a Tenax column which, after drying, was subjected to SFE. Contaminants are transferred to the capillary GC column via a length of small internal-diameter fused-silica capillary. The whole isolation process is controlled by a single ten-port valve.

7.4.2 SFE–SFC

The instrumentation (Westwood, 1992) for on-line SFE–SFC ranges from relatively simple systems to more complex arrangements involving switching valves and multiple pumps. Most systems include an extraction cell which is temperature-controlled or held in an oven, a switching valve, a cryogenically cooled trap and a chromatographic oven housing the column (usually a capillary) and a detector. The fluid may be delivered to both the cell and the column by using a single pump. When different conditions are required for simultaneous extraction and chromatography, a dual pumping system is required.

A typical SFE–SFC system is illustrated in Figure 7.11. The outlet of the cell is connected to a flow restrictor which is in turn connected to an accumulating trapping system; this may be a coated or more usually uncoated fused-silica retention gap or transfer line, or an adsorbent trap housed in a cryogenically cooled tee. During extraction, the tee is vented to the atmosphere and the extract is concentrated within the transfer line or trap. After extraction is complete, the valve is switched and supercritical fluid is introduced into the side arm of the tee to transfer the extract on to the SFC column. If uncoated fused-silica tubing is used for the retention gap, the extract is rapidly transferred to the analytical column and there concentrated by phase-ratio focusing. The process is aided by keeping the mobile phase at low density. Other more complicated systems have been reported, using off-line and multiport switching valves to allow continuous extraction or to permit venting of the extraction cell during simultaneous chromatographic analysis.

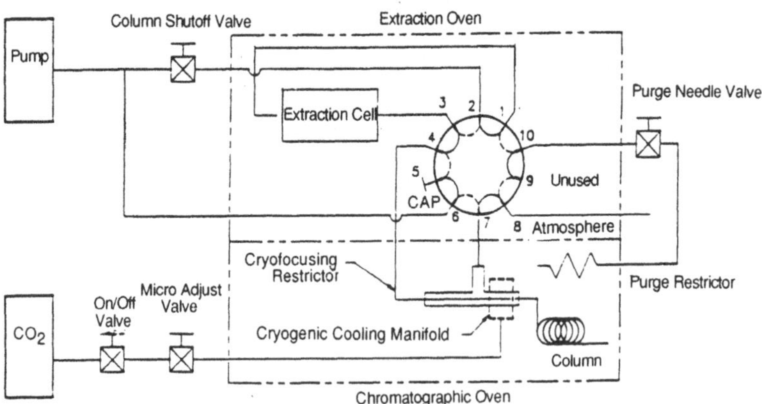

Figure 7.11 On-line coupling of supercritical fluid extraction to supercritical fluid chromatography.

7.4.3 *SFE–HPLC*

Coupled SFE–HPLC has also been described (Cortes *et al.*, 1991). For example, a system for the determination of chlorinated phenols in various solid matrices permits direct introduction of supercritical fluid extracts into an HPLC, allowing quantitative determinations down to the sub-ppm level without clean-up. The effect of adsorption on active sites in the matrix may be overcome by adding modifier to the CO_2 to increase the rate; spiked samples may be a poor guide to the necessary conditions (Figure 7.12) (Hawthorne *et al.*, 1992*a*).

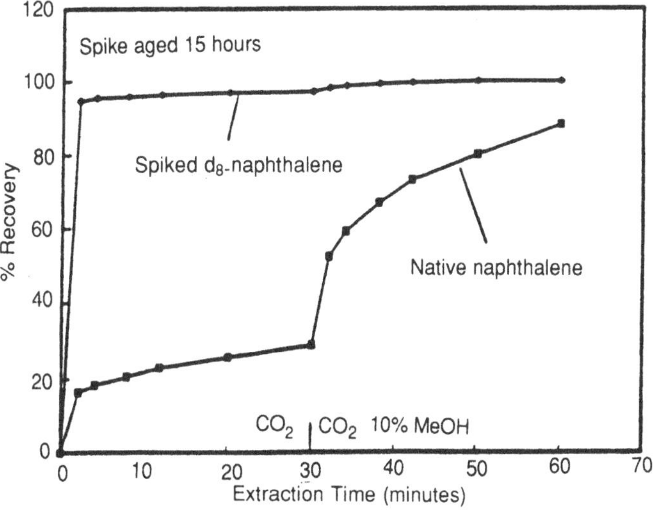

Figure 7.12 Supercritical fluid extraction recovery rates of naphthalene and spiked d_8-naphthalene from the same sample of air particulates at 400 atm and 80 °C (Hawthorne *et al.*, 1992*a*).

7.5 Applications

7.5.1 *Purgeable halocarbons*

Eleven purgeable halocarbons were isolated from a sediment matrix by on-line SFE using carbon dioxide as the supercritical fluid (Levy and Roselli, 1989). The restrictor from the SFE apparatus was directly inserted into a split/splitless injector of a gas chromatograph. The injector was kept at 250–325 °C to minimise cooling when the supercritical fluid decompressed. To focus the analytes, it was necessary to cool the gas chromatographic oven cryogenically during the SFE. The halocarbons deposited in the capillary chromatographic column were then analysed by gas chromatography–electron capture detection (GC–ECD). The SFE was performed at 250 bar and 40 °C for 10 min. The volume of the extraction cell was 2 ml. Recovery of the purgeable halocarbons from the sediment was approximately 90% at low μg g^{-1} levels.

7.5.2 Chlorobenzenes and chlorophenols

Hong-Yu (1991) described the use of off-line SFE to extract ten chlorobenzenes from sediment samples. After extraction, the analytes were preconcentrated on a C_{18} trap. Chlorobenzenes were then eluted from the trap by iso-octane and after a clean-up procedure analysed by GC–ECD. Optimum SFE conditions were as follows: extraction temperature 80 °C, extraction time 20 min, sample size 1 g placed in a 7 ml extraction chamber, extraction pressure of pure carbon dioxide 168 bar at a flow rate of liquid carbon dioxide of 2 ml min^{-1}. The method was tested using sediments with certified concentrations of chlorobenzenes and by comparisons with Soxhlet extraction. In the concentration range 5–200 ng g^{-1} for the different chlorobenzenes, SFE provided results in agreement with the certified values and/or Soxhlet extraction.

Richards and Campbell (1991) investigated the recovery of some priority pollutants from spiked soil samples by SFE, Soxhlet extraction and sonication liquid extraction. They found that the SFE was more efficient than other extraction methods for 13 of 16 compounds tested. SFE averaged 80.2% with individual values from 70.4 to 95%. Only phenolic and chlorophenolic compounds had equivalent recoveries in the SFE and Soxhlet extraction. Sonication was less efficient than the SFE for all compounds studied except 2,4-dichlorophenol, which was recovered in equal amounts by both extraction methods. Groups of compounds tested were: phenols and chlorophenols, chlorobenzenes, and naphthalene. The best results were obtained using carbon dioxide modified with 2% methanol as the supercritical fluid. The off-line extraction was performed at 395 bar and 80 °C, with 2 g of the spiked soil (spiking level 25 μg g^{-1} for each compound) extracted in to a 1.67 ml chamber for 30–40 min. In total 20 ml of the extraction fluid was consumed (measured as a liquid using pump displacement). Compounds extracted were collected at the outlet of the restrictor in methylene chloride. The methylene chloride solution was concentrated to 1 ml under a gentle stream of nitrogen and analysed by gas chromatography–mass spectrometry (GC–MS).

Recovery of chlorophenols from spiked sediment (0.5 g) was also determined off-line using SFE by Janda and Sandra (1990). The conditions used were a pressure of 230 bar at 48 °C, for 5 min with a restrictor of 20 cm long and 25 μm i.d. Analytes leaving the restrictor were trapped in 0.5 ml of methanol. After derivatisation to chlorophenol acetates, the compounds were analysed by capillary GC. Recoveries of *o*- and *p*-chlorophenol, 3,4-dichlorophenol, 2,4,5-trichlorophenol and pentachlorophenol were equal to or higher than 90%. Under the conditions described there were no differences between recoveries from dry and wet (20% of water) sediment except for *o*-chlorophenol and *p*-chlorophenol, which were partially lost during the drying.

7.5.3 Polychlorinated biphenyls

The possibility of isolation of PCBs from sediment using simple off-line SFE apparatus was described by Schantz and Chesler (1986). More recent papers deal with on-line coupled techniques, where the SFE is directly coupled usually to GC–ECD. Hawthorne and Miller (1987*a*) successfully coupled SFE to GC–ECD by inserting the restrictor outlet through the on-column injection port. The GC oven was cooled during the extraction. Extracted analytes were thermally focused inside the capillary chromatographic column at the outlet of the restrictor. Since nitrous oxide, which was used as a supercritical fluid, gives a relatively high ECD response, the restrictor was flushed for 2–3 min with carrier gas (5 °C). PCBs were then analysed by using temperature-programmed GC. Parameters of the SFE were as follows: a pressure of 300 bar at 45 °C for 10 min, and a 25 μm i.d. restrictor (corresponding to the flow rate of the gaseous nitrous oxide ± 240 ml min^{-1}) were used. This restrictor provided good extraction recoveries but required wide-bore thick-phase GC columns to provide good low-temperature focusing. The on-line coupled SFE–GC was applied to spiked (8 μg g^{-1} of Aroclor 1254) sediment (10 mg).

In the paper published by Onuska and Terry (1989*b*), PCBs extracted from sediment by the SFE were collected in an 'accumulator' (2 m × 0.32 i.d. fused-silica capillary coated with cross-linked SE-54 stationary phase). The analytes emerging from the restrictor were partitioned in the accumulator, which was kept at low temperature (5 °C). Desorption from the accumulator was achieved by a rapid temperature ramp to 140 °C using hydrogen as the carrier gas. Direct on-line coupling of the accumulator to the analytical capillary column was made by means of a six-port valve switching to another position (the first position of the six-port valve was utilised during extraction). Recommended parameters of the SFE were: supercritical fluid carbon dioxide + 2% methanol (kinetics of the SFE with pure carbon dioxide proved to be slower); pressure 207 bar at a temperature of 40 °C. The restrictor was 10 cm long, 25 μm i.d. (flow rate of the supercritical fluid 0.35 g min^{-1}). The extraction was performed in a static mode: the extraction cell was pressurised and the pressure was maintained in the cell for 2 min to reach equilibrium. Extracted analytes were then transferred to the accumulator by opening a valve in the extraction cell outlet for 30 s. The valve was then closed and the cycle was repeated five times. After the fifth step the six-port valve was switched to another position, residual methanol in the accumulator was flushed by the carrier gas at 5 °C, and GC–ECD analysis was started. From 10 to 100 mg of the sediment with a certified value of 2.02 mg kg^{-1} of PCBs were extracted and analysed. Statistical comparison of the certified and determined values revealed excellent agreement.

Recent data provided by Hawthorne *et al.* (1992a) showed that difluorochloromethane (Freon-22) is the most efficient fluid for the extraction of PCBs from sediment (most likely because of its high dipole moment in comparison with pure carbon dioxide). Methanol-modified carbon dioxide also yielded acceptable recoveries. Off-line SFE using Freon-22 was performed mostly at 100 °C and 400 bar. PCBs were trapped by inserting the restrictor outlet into several millilitres of acetone in a vial. Although the widespread use of Freons for industrial purposes is being reduced, which would affect their future analytical uses, the hydrogen-containing fluorochlorohydrocarbons have much lower influence on ozone depletion, and are more suitable for SFE.

*7.5.4 Tetrachlorodibenzo-*p*-dioxins*

Onuska and Terry (1989*a*) tested SFE for isolation of tetrachlorodibenzodioxin (2,3,7,8-TCDD) from a sediment matrix using off-line SFE. The extraction cell had a volume of 0.5 ml and 50 mg of the sediment was extracted. The sediment was spiked by the tested compound to a concentration of 0.2 μg g^{-1}. The extracts were collected in hexane, concentrated, and subsequently analysed by GC–MS. The results of the study showed that pure carbon dioxide is not able to extract tetrachlorodibenzodioxin effectively. After 30 min of extraction at 310 bar and 40 °C the recovery was only about 50%. Much better results were obtained by using pure nitrous oxide (91% recovery). The best results were provided by modification of the fluids by the addition of 2% methanol. After 30 min of extraction, the recoveries obtained by carbon dioxide + 2% methanol and by nitrous oxide + 2% methanol were 93% and 100%, respectively. Moisture content in the sediment dramatically decreased the recovery by 10–15%. To achieve the same recovery with wet sediment the time of the extraction had to be doubled. For comparison, classical Soxhlet extraction of 1 g of the sediment by hexane–acetone–trimethylpentane mixture for 18 h provided only 65% recovery of TCDD.

Supercritical fluid extraction was also used for enrichment and isolation of TCDD from municipal incinerator fly-ash (Onuska and Terry, 1991). Pure nitrous oxide with and without methanol and toluene were used. Experiments were carried out at 45 °C with 25 mg samples. The best recovery was obtained with nitrous oxide + 2% methanol at 400 bar.

Under these conditions extraction was more efficient than that of the Soxhlet technique.

Suitability of nitrous oxide for the extraction of dibenzo-*p*-dioxins and dibenzofurans from incinerator fly-ash was described by Alexandrou and Pawliszyn (1989). They found that carbon dioxide with 10% benzene as modifier was able to provide higher recoveries of these compounds than unmodified carbon dioxide. For the fractionation and clean-up of the complex organic mixtures in an organic solution obtained after the off-line SFE of the fly-ash sample, a further extraction step was used instead of the usual clean-up procedure by column liquid chromatography. This involved spiking the liquid extract on to a Florisil column. The column was then extracted for 15 min with carbon dioxide at 40 °C and 207 bar. Recoveries from the column of PCBs and chlorobenzenes were over 75%. Full recoveries of polychlorodibenzo-*p*-dioxins were obtained by extraction from the column with nitrous oxide for 90 min at 415 bar (Alexandrou *et al.*, 1992).

7.5.5 *Organochlorine pesticides*

More than twenty organochlorine pesticides were isolated from a sediment matrix by off-line SFE (Naiyer and Khidekel, 1992). From 1 to 20 g of spiked sediment were extracted at 60 °C and 250 bars for 5 min in static mode and 20 min in dynamic mode. The flow rate of the supercritical carbon dioxide was 0.5 ml min^{-1}. Methanol was used as a polarity modifier. Recovery of chlorinated pesticides under these conditions was >90%. It was also shown that extraction with carbon dioxide was inefficient for extraction of elemental sulphur from environmental solids (Nam *et al.*, 1992). Elemental sulphur is a serious problem when an electron capture detector or a flame photometric detector is used for the gas chromatographic analysis. The extracts of environmental solids after liquid extraction usually require treatment with metallic mercury or copper, which results in conversion of soluble sulphur to insoluble sulphides. The method is effective; however, the treatment leads to degradation of a number of pesticides. In contrast to Soxhlet extraction, less than 2% of the elemental sulphur was present in the SFE extracts, while the majority of pesticides from the spiked sediments were recovered.

7.5.6 sym-*Triazine herbicides*

Supercritical fluid extraction recoveries of >90% were achieved by using pure carbon dioxide for extracting propazine, terbutylazine, atrazine and cyanazine from sediment (Janda *et al.*, 1989*b*). Simazine, a very poorly soluble compound in pure carbon dioxide, required the addition of methanol to the supercritical fluid. The entrainer was added directly to the sediment in the extraction chamber (Janda *et al.*, 1989*b*). The off-line extraction was performed under the following conditions: a pressure of 230 bar at 48 °C for 30 min. The sediment (0.5 g) was spiked in the range 4 mg kg^{-1} to 40 µg kg^{-1}. Compounds leaving the restrictor from the extraction apparatus were trapped into a few millilitres of methanol. The methanolic solution was (after concentration under a gentle stream of nitrogen) analysed by capillary gas chromatography–flame ionisation detection (GC–FID) and/or high performance liquid chromatography–diode array detection (HPLC–DAD) at 225 nm. Due to a high background of natural hydrocarbons in the sediment, which were easily recovered by SFE and interfered with *sym*-triazines during the GC analysis, the HPLC method was found to be more suitable for the final analysis. It was also confirmed in another study with actual soil samples (Tehrani *et al.*, 1992), that the presence of methanol or acetone in the supercritical carbon dioxide improved the recovery of atrazine.

7.5.7 *Phenoxycarboxylic acid herbicides*

Experiments to determine the solubility of some pesticides in supercritical carbon dioxide (Schafer and Baumann, 1988) and model experiments determining the recovery from glass wool (Schafer and Baumann, 1989) showed the potential of SFE for environmental samples containing phenoxycarboxylic acids and derivatives.

Soil spiked with 1.8–7.5 µg g^{-1} of the iso-octyl ester of 2,4,5-trichlorophenoxyacetic acid was extracted off-line under the following conditions (Firor, 1990): pressure of 100–316 bar at temperatures of 40–45 °C, with an extraction time of 40 min and a flow rate of the liquid carbon dioxide of 2–2.5 ml min^{-1}. The soil sample (10 g) was mixed prior to extraction with water and methanol, 5% and 3% by weight, respectively, in order to increase the polarity of the supercritical fluid. It was shown that the higher the pressure, the higher the recovery of the herbicide. However, at the higest pressure the recovery did not exceed 80%.

7.5.8 *Fuels and crude oil*

Fuel and crude oil contamination of solids can be easily isolated by SFE using pure carbon dioxide. Hawthorne *et al.* (1992*b*) showed that recovery of C_5–C_{40} *n*-hydrocarbons from spiked alumina approached 100% (carbon dioxide at 325 bar, temperature 60 °C and time 20 min). The method was also used for the determination of biomarkers (pristane and phytane) and carbon number distribution in crude oil source rocks (extraction pressure 400 bar for a 20 min extraction).

In addition, Hawthorne *et al.* (1990) discovered that, from a wet sediment (20% water) contaminated by fuel, C_8–C_{30} *n*-hydrocarbons, alkylbenzene and alkylnaphthalene isomers could be removed from the sample (1.3 g) by pure carbon dioxide (380 bar and 50 °C) in 10 min. It was important that the SFE proved efficient extraction from wet samples, because lower *n*-alkanes are lost during the drying procedure. Quantitative analysis was performed using on-line SFE-capillary GC–MS with a conventional split/splitless injector. The SFE–GC coupling was achieved by inserting the restrictor (10 cm long, 25 μm i.d.) directly into the split/splitless injector.

Pure supercritical carbon dioxide was also found to be an excellent fluid for extraction of diesel fuel from clay and soil (Messer and Taylor, 1992; Emery *et al.*, 1992*b*). A range of hydrocarbons from C_{14} to C_{20} was monitored. The off-line SFE was performed at 314 bar and 70 °C. Matrix sample size was 1 g and flow rate of the liquid carbon dioxide was 2 ml min^{-1}. Up to 100 volumes of the SFE cartridge of the liquid carbon dioxide were used for the SFE. The analytes extracted were then trapped on to a small C_{18} column prior to elution with 3 ml of methylene chloride. The eluate was analysed by GC–FID.

7.5.9 *Polycyclic aromatic hydrocarbons*

PAHs are one of the most troublesome groups of compounds, because of their poor solubility in supercritical fluids (especially compounds with more condensed benzene rings). The recovery is also strongly dependent on the type of matrix. Recovery of PAHs usually decreases with increasing molecular weight (decreasing solubility). In matrices such as fly-ash and sediments they are adsorbed very strongly and their complete recovery by a supercritical fluid is sometimes difficult. It is particularly important to consider the type of matrix to be analysed for PAHs prior to analysis. For example, it was shown by Wright *et al.* (1987*a*), that PAHs are not totally recovered from urban dust by carbon dioxide at 80 bar (density 0.23 g ml^{-1}), whilst extraction from spiked glass beads gave higher recoveries.

The different solubilities of various groups of hydrocarbons in supercritical fluids can be used for removal of compounds that would otherwise interfere during final chromatographic analysis. Thus SFE can provide a certain degree of selectivity. It was shown by Hawthorne and Miller (1986) that prevailing portions of *n*-alkanes (nonadecane–hexacosane) were removed from diesel exhaust particulates by carbon dioxide at 75 bar, while PAHs were not significantly extracted. The PAHs were then extracted at 300 bar, with recovery mostly above 90%.

There has not yet been established an unambiguous analytical scheme for the extraction of PAHs from various matrices. However, probably the best results were obtained with nitrous oxide + 5% methanol as supercritical fluid. Recoveries of PAHs using this fluid were much higher than recoveries obtained with carbon dioxide or ethane, and even higher than with carbon dioxide + 5% methanol (Hawthorne and Miller, 1987). This was probably due to the higher polarity of the nitrous oxide + methanol and resultant higher solubility of the PAHs in this fluid. Typical conditions used for the extraction of PAHs from various matrices are given in Table 7.5.

A promising supercritical fluid for the extraction of PAHs is difluorochloromethane (Freon-22) (Hawthorne *et al.*, 1992*a*). It has been recently shown that extraction rates of individual PAHs from a petroleum sludge are similar with Freon-22, whilst the rates decreased with increasing molecular weight using carbon dioxide and nitrous oxide. Extraction with difluorochloromethane for 40 min was much more efficient than methylene chloride sonication for 18 h, especially for higher PAHs. Trifluoromethane also provided good recoveries of PAHs from soil (Emery *et al.*, 1992*a*).

7.5.10 Linear alkylbenzenesulphonates

Quantitative extraction of anionic surfactants from soil, sediment, and municipal waste-water treatment sludge has been achieved by using supercritical carbon dioxide, with polarity increased by a high content (about 40 mol%) of methanol (Hawthorne *et al.*, 1991). Recovery of alkylbenzenesulphonates was higher than 90% after 30 min at 380 bar and 125 °C. The amount of the sample to be extracted was usually 1 g. The flow of the extraction fluid through the extraction cartridge was 1.2 or 0.45 ml min^{-1}, using 10 cm lengths of either 30 or 25 μm i.d. fused-silica tubing as a restrictor. The compounds extracted were collected in vials containing 5 ml of ethanol. For the recovery measurement ^{14}C-labelled compounds were used. The native alkylbenzenesulphonates were analysed by HPLC with fluorescence detection. The off-line apparatus used comprised a simple vessel for the preparation of a high content of polarity modifier in the supercritical fluid.

7.5.11 SFE of organic compounds from adsorbent materials

SFE can be used for the extraction of organic compounds from Tenax, XAD resins, reversed-phase based sorbents, charcoal or polyurethane foams after previous enrichment of oganic compounds from environmental samples (e.g. water, air, etc.).

Hawthorne and Miller (1986) showed that PAHs from diesel exhausts could be effectively removed from Tenax (80 mg) by carbon dioxide under the following conditions: 45 °C and 200 bar for 15 min. Naphthalene, 9-fluorenone, phenanthrene, pyrene, benzo[*a*]anthracene, benzo[*ghi*]perylene and coronene provide recoveries higher than 90%. Similar results were obtained for XAD-2 resin and polyurethane foam (Wright *et al.*,

Table 7.5 Supercritical fluid extraction of PAHs from environmental solids

Matrix, amount	Fluid	Time of SFE	Temperature (°C)	Pressure (MPa)	Concentration order	SFE set-up	Reference
Carbon black, 450 mg	CO_2	Static 5 min	50	32.8	μg/g	On-line	Levy and Rosselli (1989)
Urban dust, 6 g	CO_2	4 h	40	36.2	μg/g	Off-line	Schantz and Chessler (1986)
Sediment, 10 mg	N_2O	10 min	45	30.6	μg/g	On-line	Hawthorne and Miller (1987*a*)
Urban dust, 15 ml SFE cell	CO_2	Static and 1 min dynamic	50	20	μg/g	On-line	Wright *et al.* (1987*a*)
Dust, 5–30 mg	CO_2	90 min	45	30.6	μg/g	Off-line	Hawthorne and Miller (1986)
	CO_2 + CH_3OH	30 min	65	30.6	μg/g	Off-line	Hawthorne and Miller (1986)
Dust, sediment, fly ash, 20–50 mg	Ethane, N_2O, CO_2, N_2O + CH_3OH	30 min	45–65	30.6	μg/g	On-line	Hawthorne and Miller (1987*b*)
Petroleum waste sludge, 310 mg	$CHClF_2$	40 min	100	40	μg/g	Off-line	Hawthorne *et al.* (1992*a*)

1987*b*). On the other hand, PAHs were not recovered from Spherocarb. However, Soxhlet extraction was also unsuccessful in this case (Wright *et al.*, 1987*b*).

Supercritical fluid extraction was also used for determination of semi-volatile mutagens in air using solid adsorbents (Wong *et al.*, 1991). Adsorbents (finally XAD-4 resin was utilised) which were used for the trapping of the mutagens from air were extracted by pure and by entrainer-modified carbon dioxide. Mutagens, 4-nitrobiphenyl, 2-nitrofluorene and fluoranthene, were recovered from a XAD-4 trap in 180 min with efficiencies of 88.5, 92.3 and 60.6%, respectively, using carbon dioxide with 12% of hexane at 415 bar and 50 °C (Wong *et al.*, 1991).

Chlorinated benzenes (tetrachlorobenzene, pentachlorobenzene and hexachlorobenzene) were isolated from water ($1\,\mu g\,l^{-1}$) by solid-phase extraction (Amberlite XAD resins, Tenax GC and HPLC C_{18} reversed-phase). The analytes were, after drying of the sorbent by means of a stream of nitrogen at 60 °C, released from the adsorbent material by supercritical carbon dioxide. In the work, Pawlyszin and Alexandrou (1989) used carbon dioxide with 10% of methanol at 400 bar and 60 °C for 45 min. The restrictor used was a piece of 20 μm i.d. fused-silica capillary. Overall recovery of chlorobenzenes was >97%.

Sulphonyl urea herbicides were isolated from a water sample by solid-phase extraction by Howard *et al.* (1992). The analytes were then eluted from the extraction disk by SFE using carbon dioxide + 5% methanol. For the final analysis HPLC–UV (235 nm) was used. Compounds tested were: sulphachloropyridazine, thifensulphuron methyl, metsulphuron methyl, sulphometuron methyl, chlorsulphuron, tribenuron methyl, benzsulphuron methyl and chlorimuron ethyl. Recovery of the herbicides was mostly >80%. Only tribenuron methyl and chlorimuron ethyl had recoveries lower than 80%. Recoveries were measured from a 1 l water sample at a concentration level of $50\,\mu g\,l^{-1}$.

Solid-phase extraction with SFE elution was also used for the analysis of explosives in water at the $\mu g\,l^{-1}$ to $ng\,l^{-1}$ level (Slack *et al.*, 1992). 2,6-Dinitrotoluene, 2,4-dinitrotoluene, and trinitrotoluene were adsorbed on a phenyl stationary phase. The nitrotoluenes were eluted by supercritical carbon dioxide. Before SFE (400 bar and 75 °C), the sorbent was doped with toluene. On-line SPE–SFE–GC–ECD was used for analysis of the water samples. The method gave recoveries of approximately 100%.

7.5.12 Direct SFE of organics from water samples

There are several problems associated with the extraction of compounds from aqueous solution. Probably the main problem is the relatively high solubility of water in the supercritical carbon dioxide, approximately 0.3%

(Kuk and Montagna, 1983). For dynamic extraction, which is commonly used for the extraction of solids, the removal of the water phase and enrichment of the supercritical fluid by water can cause 'principal' (the water phase is transferred through the restrictor into the collection vessel or accumulator at off-line methods, and/or would enter the chromatographic system, which is not desirable in most cases) and technical (plugging of the restrictor by ice during the supercritical fluid expansion) problems. The problems have been recently partly solved, but the present state-of-the-art of the SFE of organics from water is not definitely as acceptable for routine analysis as the SFE of environmental solids.

The first papers about the SFE of water samples appeared in the late eighties (Hedrick and Taylor, 1989; Thiebaut *et al.*, 1989). The first method (Hedrick and Taylor, 1989) was based on the 'closed loop stripping' principle, where the supercritical fluid was (after pressurising of the system) recycled by a pump from the outlet of the extraction cell back into the water sample. After equilibration in the whole system was achieved, a sample of the supercritical phase was taken by means of a valve with loop. The content of the loop was then analysed by supercritical fluid chromatography.

The system described above was tested for the analysis of di-isopropyl methylphosphonate, at a concentration of 834 $\mu g\ l^{-1}$ to 834 $mg\ l^{-1}$, in water. The volume of the water sample was 8 ml. The time necessary to reach equilibrium was 1.5 h. When 0.1 mg of NaCl was added to the water sample, prior to extraction, the equilibration time was reduced to less than 5 min. The relative standard deviation was 15% for a concentration of 834 $\mu g\ l^{-1}$ and 1.5% for concentrations higher than 8.34 $mg\ l^{-1}$. The method was also used for the extraction of phenol from aqueous solutions (Hedrick and Taylor, 1990).

The other system (Thiebaut *et al.*, 1989) was based on a sandwich-type phase separator, in which the supercritical carbon dioxide and water phase, after passing through the extraction coil, are separated by means of a hydrophobic membrane. Two membranes were able to withstand higher pressure necessary: PVDF $[(-CH_2-CF_2-)_n]$ and Delrin $[(-CH_2-O-)_n]$. A sample of the separated supercritical fluid was taken by a valve with loop for SFC analysis. Phenol and 4-chlorophenol were utilised as test compounds.

Water solutions can also be extracted by a supercritical fluid when a smaller amount of water is added to the inert material (sand, glass beads, etc.). The film of the water phase on the surface of the carrier material can be subjected to SFE. In this case the restrictor has to be attached in the upper part of the extraction chamber, to prevent liquid water from entering the restrictor. This method was used for the SFE of phenol from water (Hawthorne *et al.*, 1992*b*).

7.5.13 *Simultaneous SFE and derivatisation*

The *in situ* derivatisation of polar compounds is quite challenging. Under extraction conditions (especially at higher pressures) the derivatisation reactions of highly polar compounds could easily take place (Hills *et al.*, 1991; Miller *et al.*, 1991), resulting in less-polar products, which are more suitable for extraction and subsequent chromatographic analysis.

A series of interesting papers has recently been published (Hawthorne *et al.*, 1992*b*; Miller and Hawthorne, 1992). For the derivatisation of polar compounds (2,4-dichlorophenoxyacetic acid and Dicamba, phospholipid fatty acids in whole cells and phenolics in waste water) trimethylphenylammonium hydroxide and boron trifluoride in methanol were used. The procedure consisted of four steps: (1) the sample was placed in the extraction cell together with the derivatising reagent; (2) the cell was placed into the heater and pressurised by carbon dioxide to 400 bar; (3) derivatisation took place for 5–45 min under static conditions (the outlet of the extraction cell was closed); and finally (4) the outlet of the cell was opened and the sample was extracted dynamically for 5–15 min. Derivatised compounds—methyl esters and/or anisols—were trapped at the outlet of the restrictor into small amounts of methanol or dichloromethane. The solutions were analysed by GC–FID, -ECD and -MS. Recovery for all compounds was >90%.

In situ derivatisation–SFE was also used for acetylation of phenols isolated by solid-phase extraction from water samples (Tang and Ho, 1992). After the pH value of the water sample was adjusted to 12 it was passed through the conditioned anion exchange disk. Anionic forms of the phenols were thus trapped in the exchanger. After solid-phase extraction, 500 μl of acetic anhydride was added and the derivatisation to the acetylated phenols took place under static SFE conditions (400 bar and 50 °C). After derivatisation, the phenolic acetates were eluted with 30 ml of supercritical carbon dioxide. The SFE extract was trapped in 2 ml of acetone and analysed by GC–MS. Recovery of all phenols tested (2- and 4-nitrophenol and 1-naphthol) was higher than 75% of the 25–50 $\mu g\ l^{-1}$ level. The total time for the derivatisation–SFE was about 30 min.

References

Alexandrou, N. and Pawliszyn, J., (1989), Supercritical fluid extraction for the rapid determination of polychlorinated dibenzo-*p*-dioxins and dibenzofurans in municipal incinerator fly ash. *Anal. Chem.*, **61**, 2770–2776.

Alexandrou, N., Lawrence, M.J. and Pawliszyn, J., (1992), Cleanup of complex organic mixtures using supercritical fluids and selective adsorbents. *Anal. Chem.*, **64**, 301–311.

Ashraf, S., Bartle, K.D., Clifford, A.A., Moulder, R., Raynor, M.W. and Shilstone, G.F., (1992), Prediction of the conditions for supercritical fluid extraction of atrazine from soil. *Analyst*, **117**, 1697–1700.

Bartle, K.D., Clifford, A.A. and Shilstone, G.F., (1989), Prediction of solubilities for tar extraction by supercritical carbon dioxide. *J. Supercritical Fluids*, **2**, 30–34.

Bartle, K.D., Clifford, A.A., Hawthorne, S.B., Langenfeld, J.J., Miller, D.J. and Robinson, R.E., (1990*a*), A model for dynamic extraction using a supercritical fluid. *J. Supercritical Fluids*, **3**, 143–149.

Bartle, K.D., Clifford, A.A., Jafar, S.A., Kithinji, J.P. and Shilstone, G.F., (1990*b*), Use of chromatographic retention measurements to obtain solubilities in a liquid or supercritical mobile phase. *J. Chromatogr.*, **517**, 459–476.

Bartle, K.D., Clifford, A.A., Jafar, S.A. and Shilstone, G.F., (1991), Solubilities of solids and liquids of low volatility in supercritical carbon dioxide. *J. Phys. Chem. Ref. Data*, **20**, 713–756.

Bartle, K.D., Boddington, T., Clifford, A.A. and Hawthorne, S.B., (1992), The effect of solubility on the kinetics of dynamic supercritical fluid extraction. *J. Supercritical Fluids*, **5**, 207–212.

Cortes, H.J., Green, L.S. and Campbell, R.M., (1991), On-line coupling of supercritical fluid extraction with multidimensional microcolumn liquid chromatography/gas chromatography. *Anal. Chem.*, **63**, 2719–2724.

Emery, A., Chesler, S. and MacCrehan, W., (1992*a*), Trifluoromethane (CHF_3) for the SFE of environmental samples. *Abstracts of the 4th International Symposium on Supercritical Fluid Chromatography and Extraction*, Cincinnati, OH, USA, 19–22 May, p. 45.

Emery, AP., Chesler, S.N., MacCrehan, W.A. and Yoder, R.E., (1992*b*), Recovery of environmental analytes from clays and soils by SFE/GC. *Abstracts of the 4th International Symposium on Supercritical Fluid Chromatography and Extraction*, Cincinnati, OH, USA, 19–22 May, p. 225.

Firor, R.L., (1990), *Electron Capture Detector Evaluation of Soil Extracts Containing 2,4,5-Trichlorophenoxyacetic Acid*. Application Note 228–113, Hewlett-Packard, Avondale, PA, USA.

Furton, KG. and Rein, J., (1992), The quantitative effect of microextractor cell geometry on the analytical supercritical fluid extraction efficiencies of environmentally important components. *Chromatographia*, **31**, 297–299.

Hawthorne, S.B., (1990), Analytical scale supercritical fluid extraction. *Anal. Chem.*, **62**, 633A–642A.

Hawthorne, S.B. and Miller, D.J., (1986), Extraction and recovery of organic pollutants from environmental solids and Tenax–GC using supercritical carbon dioxide. *J. Chromatogr. Sci.*, **24**, 258–263.

Hawthorne, S.B. and Miller, D.J., (1987*a*), Directly coupled supercritical fluid extraction–gas chromatographic analysis of polycyclic aromatic hydrocarbons and polychlorinated biphenyls from environmental solids. *J. Chromatogr.*, **403**, 63–76.

Hawthorne, S.B. and Miller, D.J., (1987*b*), Extraction and recovery of polycyclic aromatic hydrocarbons from environmental solids using supercritical fluids. *Anal. Chem.*, **59**, 1705–1708.

Hawthorne, S.B., Miller, D.J. and Langenfeld, J.J., (1990), Quantitative analysis using directly coupled supercritical fluid extraction–capillary gas chromatography (SFE–GC) with a conventional split/splitles injection port. *J. Chromatogr. Sci.*, **28**, 2–8.

Hawthorne, S.B., Miller, D.J., Walker, D.D., Whittington, D.E. and Moore, B.L., (1991), Quantitative extraction of linear alkylbenzenesulphonates using supercritical carbon dioxide and a simple device for adding modifiers. *J. Chromatogr.*, **541**, 185–194.

Hawthorne, S.B., Langenfeld, J.J., Miller, D.J. and Burford, M.D., (1992*a*), Comparison of supercritical $CHClF_2$, N_2O and CO_2 for the extraction of polychlorinated biphenyls and polycyclic hydrocarbons. *Anal. Chem.*, **64**, 1614–1622.

Hawthorne, S.B., Miller, D.J., Nivens, D.E. and White, D.C., (1992*b*), Supercritical fluid extraction of polar analytes using *in-situ* chemical derivatization. *Anal. Chem.*, **64**, 405–412.

Hedrick, J. and Taylor, L.T., (1989), Quantitative supercritical fluid extraction and chromatography of a phosphonate from aqueous media. *Anal. Chem.*, **61**, 1986–1988.

Hedrick, J.L. and Taylor, L.T., (1990), Supercritical fluid extraction strategies of aqueous based matrices. *J. High Resolut. Chromatogr.*, **13**, 312–316.

Hills, J.W., Hill, H.H. and Maeda, T., (1991), Simultaneous supercritical fluid derivatization and extraction. *Anal. Chem.*, **63**, 2152–2155.

Hong-Yu, R., (1991), *Water Technology Centre Newsletter* No. 21, Environment Canada, Burlington, Canada.

Howard, A.L., Yost, K.J. and Taylor, L.T., (1992), Quantitative SFE of sulfonyl urea herbicides from water. *Abstracts of the 4th International Symposium on Supercritical Fluid Chromatography and Extraction*, Cincinnati, OH, USA, 19–22 May, p. 141.

Janda, V. and Sandra, P., (1990), Extraction of chlorophenols from sediment by supercritical carbon dioxide. *Hydrochémia*, **90**, 295–300.

Janda, V., Steenbeke, G. and Sandra, P., (1989*a*), In *Proceedings of the 10th International Symposium on Capillary Chromatography*, Sandra, P. and Redant, G. (Eds), Dr Alfred Hüthig Verlag, Heidelberg, Germany, p. 457.

Janda, V., Steenbeke, G. and Sandra, P., (1989*b*), Supercritical fluid extraction of *S*-triazine herbicides from sediment. *J. Chromatogr.*, **479**, 200–205.

King, J.W., (1990), Applications of capillary supercritical fluid chromatography–supercritical fluid extraction to natural products. *J. Chromatogr. Sci.*, **28**, 9–14.

Kuk, M.S. and Montagna, J.C., (1983), In *Chemical Engineering at Supercritical Fluid Conditions*, Paulitis, M.E., Penninger, J.M., Gray, R.D. and Davidson, K.P. (Eds), Ann Arbor Science, Ann Arbor, MI, USA, p. 101.

Lee, M.L. and Markides, K.E., (1990), *Analytical Supercritical Fluid Chromatography and Extraction*, Chromatography Conferences Inc., Provo, UT, USA.

Levy, J.M. and Rosselli, A.C., (1989), Quantitative supercritical fluid extraction coupled to capillary gas chromatography. *Chromatographia*, **28**, 613–616.

Liu, Z., Farnsworth, P.B. and Lee, M.L., (1992), High speed thermally modulated SFE/GC for the analysis of volatile organic compounds in solid matrices. *J. Microcol. Sep.*, **4**, 199–208.

McHugh, M. and Krukonis, V., (1986), *Supercritical Fluid Extraction. Principles and Practice*, Butterworth, Boston, MA, USA.

Messer, D.C. and Taylor, L.T., (1992), SFE of diesel fuel from clay and soil matrices. *Abstracts of the 4th International Symposium on Supercritical Fluid Chromatography and Extraction*, Cincinnati, OH, USA, 19–22 May, p. 137.

Miller, D.J. and Hawthorne, S.B., (1992), Chemical derivatization/SFE of polar and ionic analytes. *Abstracts of the 4th International Symposium on Supercritical Fluid Chromatography and Extraction*, Cincinnati, OH, USA, 19–22 May, p. 55.

Miller, D.J., Hawthorne, S.B. and Langenfeld, J.J., (1991), In *Proceedings of the 13th International Symposium on Capillary Chromatography*, Sandra, P. (Ed.), Dr Alfred Hüthig Verlag, Heidelberg, Germany, p. 477.

Naiyer, J. and Khidekel, R., (1992), Recovery of organochlorine pesticides and PCBs from environmental sediments using SFE technique. *Abstracts of the 4th International Symposium on Supercritical Fluid Chromatography and Extraction*, Cincinnati, OH, USA, 19–22 May, p. 131.

Nam, K.S., Kapila, S. and Puri, R.K., (1992), Evaluation of SFE for eliminating sulphur interference during the determination of multiresidue pesticides from sediments. *Abstracts of the 4th International Symposium on Supercritical Fluid Chromatography and Extraction*, Cincinnati, OH, USA, 19–22 May, p. 129.

Onuska, F.I. and Terry, K.A., (1989*a*), Supercritical fluid extraction of 2,3,7,8-tetrachlorodibenzo-*p*-dioxin from sediment samples. *J. High Resolut. Chromatogr.*, **12**, 357–361.

Onuska, F.I. and Terry, K.A., (1989*b*), Supercritical fluid extraction of PCBs in tandem with high resolution gas chromatography in environmental analysis. *J. High Resolut. Chromatogr.*, **12**, 527–531.

Onuska, F.I. and Terry, K.A., (1991), Supercritical fluid extraction of polychlorinated dibenzo-*p*-dioxins from municipal incinerator fly ash. *J. High Resolut. Chromatogr.*, **14**, 829–834.

Page, S.H., Sumpter, S.R. and Lee, M.L., (1992), Fluid phase equilibria in SFC with CO_2-based mobile phases: A review. *J. Microcol. Sep.*, **4**, 91–122.

Pawlyszin, J.B. and Alexandrou, N., (1989), Indirect supercritical fluid extraction of organics from water matrix samples. *Water Poll Res. J. Canada*, **24**, 207–214.

Pyle, S.M. and Setty, M.M., (1991), Supercritical fluid extraction of high sulphur soils, with use of a copper scavenger. *Talanta*, **28**, 1125–1128.

Raynor, M.W., Davies, I.L., Bartle, K.D., Clifford, A.A., Williams, A., Chalmers, J.W. and Cook, B.W., (1988), Supercritical fluid extraction/capillary supercritical fluid chromatography/Fourier transform infrared microspectrometry of polycyclic aromatic compounds in a coal tar pitch. *J. High Resolut. Chromatogr.*, **11**, 766–775.

Richards, M. and Campbell, R.M., (1991), Comparison of supercritical fluid extraction, Soxhlet and sonication methods for the determination of priority pollutants in soil. *Liq. Chromatogr. Gas Chromatogr. Intl.*, **4**, 33–36.

Schafer, K. and Baumann, W., (1988), Solubility of some pesticides in supercritical carbon dioxide. *Fresenius Z. Anal. Chem.*, **332**, 122–124.

Schafer, K. and Baumann, W., (1989), Supercritical fluid extraction of pesticides. I. Extraction properties of selected pesticides in carbon dioxide. *Fresenius Z. Anal. Chem.*, **332**, 884–889.

Schantz, M.M. and Chesler, N., (1986), Supercritical fluid extraction procedure for the removal of trace organics species from solid samples. *J. Chromatogr.*, **363**, 397–401.

Shilstone, G.F., Raynor, M.W., Bartle, K.D., Clifford, A.A., Davies, I.L. and Jafar, S.A., (1990), Coupled supercritical fluid extraction/chromatography of polycyclic hydrocarbons: Correlation with solubility calculations. *Polycyclic Aromatic Compounds*, **1**, 99–108.

Slack, G.C., McNair, H.M., Hawthorne, S.B. and Miller, D.J., (1992), *Abstracts of the 4th International Symposium on Supercritical Fluid Chromatography and Extraction*, Cincinnati, OH, USA, 19–22 May, p. 157.

Tang, P.H. and Ho, J.S. (1992), SFE of phenolics from water via liquid–solid extraction disk and on-disk derivatization. *Abstracts of the 4th International Symposium on Supercritical Fluid Chromatography and Extraction*, Cincinnati, OH, USA, 19–22 May, p. 211.

Tehrani, H., Myer, L., Damian, J., Lieschevski, P. and Algaier, J., (1992), Parameters affecting analyte recovery in SFE—Environmental and industrial applications. *Abstracts of the 4th International Symposium on Supercritical Fluid Chromatography and Extraction*, Cincinnati, OH, USA, 19–22 May, p. 143.

Thiebaut, D., Chervet, J.P., Vannoort, R.W., De Jong, G.J., Brinkman, U.A.Th., and Frei, R.W., (1989), Supercritical fluid extraction of aqueous samples and on-line coupling to supercritical fluid chromatography. *J. Chromatogr.*, **477**, 151–159.

Westwood, S.A. (Ed.), (1992), *Supercritical Fluid Extraction and its Use in Chromatographic Sample Preparation*, Blackie Academic and Professional, Glasgow, UK.

Wong, J.M., Kado, N.Y., Kuzmicky, P.A., Ning, H.-S., Woodrow, J.E., Hsich, D.P.H. and Seiber, J.N., (1991), Determination of volatile and semivolatile mutagens in air using solid adsorbents and supercritical fluid extraction. *Anal. Chem.*, **63**, 1644–1650.

Wright, B.W., Frye, S.R., McMinn, D.G. and Smith, R.D., (1987*a*), On-line supercritical fluid extraction–capillary gas chromatography. *Anal. Chem.*, **59**, 640–644.

Wright, B.W., Wright, C.W., Gale, R.W. and Smith, R.D., (1987*b*), Analytical supercritical fluid extraction of adsorbent materials. *Anal. Chem.*, **59**, 38–44.

8 Future developments in analytical supercritical fluid technology: some ideas from recent research

T.P. LYNCH

8.1 Introduction

The future development of analytical supercritical fluid technology will be guided by many factors including the research efforts of instrument manufacturers, industrialists and academics. In this final chapter we shall attempt to pose some questions where research seems to be lacking and summarise some of the current directions of research into supercritical fluids. This will hopefully seed some ideas for new research and highlight areas where this research will show application benefits. The areas can be grouped under three main sections: Theory and Understanding, Instrumentation, and Special Techniques. However, it will soon become apparent to the reader that the sections are complementary as any new special techniques require an understanding of theory and the necessary instrumentation to make them work.

8.2 Theory and understanding

Unfortunately, as in many other areas of science, analytical chemists have a great tendency to 'jump on the bandwagon' when a new technique comes along before they really understand how it works. This has certainly been the case with supercritical fluid extraction (SFE), and to a lesser extent with supercritical fluid chromatography (SFC), as chromatographic theory developed for liquid chromatography (LC) and gas chromatography (GC) has been applied fairly successfully to SFC. The situation can be further compounded by instrument manufacturers who cannot wait to supply their customers with their latest state-of-the-art equipment and by analysts who demand multisample automated equipment before the fundamentals of the technique have been fully understood. This can result in the premature rejection of a technique as customers find that the instrumentation they have purchased does not deliver the ultimate solutions indicated by the promotional literature. Therefore, there is a real need for a better understanding of the processes occurring in the supercritical state and the fundamental interactions between solutes and fluids. Furthermore, this knowledge needs to be translated into rules and methodologies that can be easily understood and applied by practising analysts.

Since the classical work of Francis (1954), who in a single paper produced solubility data for 261 substances in near-critical carbon dioxide and phase diagrams for 464 ternary systems, there has been a wealth of solubility data published for solutes in near- and supercritical carbon dioxide. However, these data cover primarily binary systems, and there are very few of those which pose real-life problems for the analyst. The problem is that when we begin working near or above the critical point the theories of solution physical chemistry that have been developed over the last hundred years or so just cannot explain the often dramatic effects that have been observed.

In recent years the growing interest in supercritical fluids for process applications has stimulated new fundamental research into the development of theoretical and predictive models for supercritical systems. This has been reflected by the publication of two comprehensive texts (McHugh and Krukonis, 1986; Bruno and Ely, 1991) on the subject, and the introduction in 1988 of the *Journal of Supercritical Fluids*, which is dedicated to the theory and application of this fascinating subject. All of these are recommended as reference sources.

In terms of computer-based application software, a commercial package is now available, SF-Solver (1991), for calculating and graphically displaying the parameters useful for working with pure and modified supercritical fluids. The features available include:

- Pressure versus density isotherms of pure and binary supercritical fluids
- Critical parameter and acentric factor calculation for a binary fluid
- Hildebrand solubility parameter calculation for pure and binary supercritical fluids
- Searchable, user-expandable libraries for solubility and critical parameter data

Figure 8.1 shows a typical printout for a set of pressure versus density isotherms for 10 mol% methanol in carbon dioxide. The data reported include the plotted isotherms, the critical parameters, and the Hildebrand solubility index. The numerical values for all curves can be obtained by moving the cursor across the screen for selected pressures. Figure 8.2 shows a typical solubility-type plot with the solubility of biphenyl in carbon dioxide plotted at three different temperatures. This type of tool is just what the analyst needs to help develop and apply analytical supercritical fluid methodology, and we look forward to future developments and to updated and shared databases.

Bartle *et al.* (1990) have made extensive studies on the kinetics of extraction and have successfully employed the 'hot-ball' method to develop

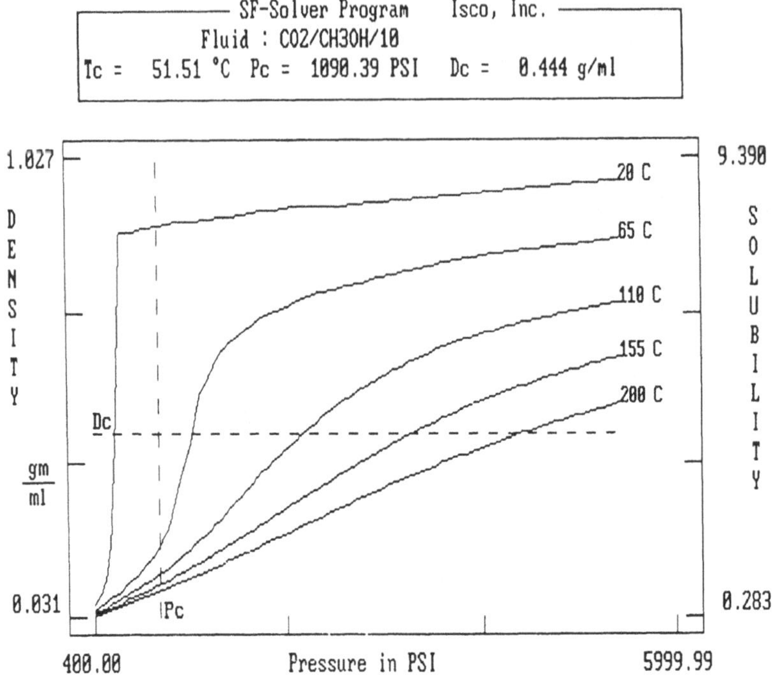

Figure 8.1 Set of pressure versus density isotherms for 10 mol% methanol in carbon dioxide obtained from SF-Solver™ software.

a model to predict the extraction profile with time. The model has been shown to be valid for a number of real systems and has been applied to the extraction of polymer additives from beads and, using the geometry of the infinite slab, polymer films. This model has recently been applied to great effect by Liu *et al.* (1992) who have developed a high-speed, thermally modulated SFE–GC which permits simultaneous sample extraction and analysis. Samples of extract were introduced to the high-speed GC at 10 s intervals, where they were separated to give a set of chromatograms while the extraction proceeded. With this system the extract was sampled continuously, thereby allowing the extraction to be monitored in almost real time for selected components. A schematic diagram of the system is shown in Figure 8.3. The performance of the system was studied by the analysis of a soil sample spiked with polycyclic aromatic hydrocarbons (PAHs), and a set of sampled high-speed chromatograms (collected 40–80 s after the start of the extraction) are shown in Figure 8.4. Four consecutive chromatograms, featuring three components each (indene, naphthalene and acenaphthene) are shown in Figure 8.4, and, by comparison of peak areas, it can be seen that during this time period the concentration of both indene and naphthalene in the extract decreased whereas the concentration of

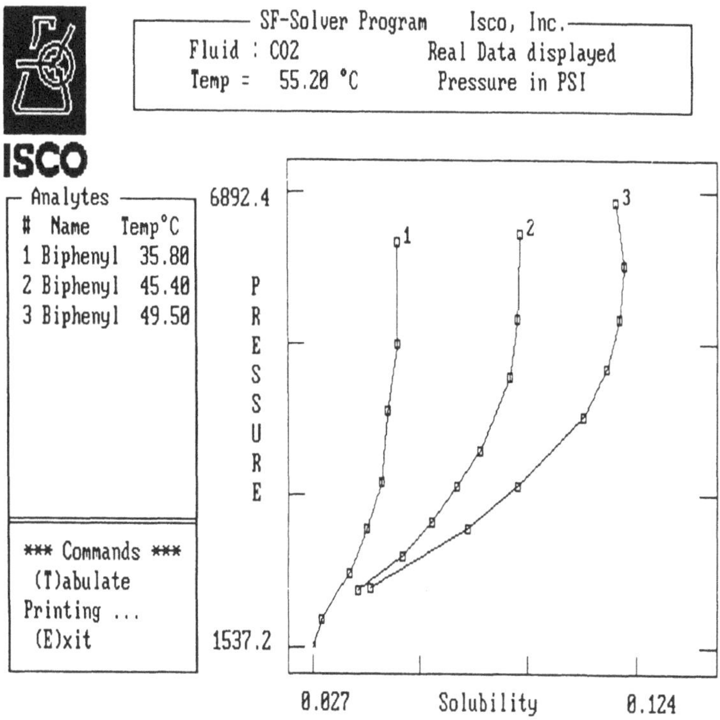

Figure 8.2 Solubility plots for biphenyl in carbon dioxide at three different temperatures as produced by the SF-Solver™ software.

acenaphthene increased. This type of system can give a lot of data in a very short time and should therefore be a useful tool in designing and optimising extraction processes.

Figure 8.5 shows a three-dimensional representation of an SFE–GC analysis of the spiked soil sample obtained using the system. In this representation one axis gives the retention times of the components in the high-speed chromatograms and the other axis gives the extraction time. In this case, 120 10 s chromatograms were generated during the 20 min extraction period.

Quantitative results were obtained using the Bartle–Clifford model by extrapolation of curves obtained by signal averaging of segmented chromatograms from the collected set. This meant that extractions did not have to be taken to completion and therefore significant time savings could be made. This type of approach warrants further study and could be a routine tool in the near future. Another related area of research that could be applied by practising analysts is the development of statistical experimental design and optimisation procedures for separation processes.

Huang and Pawliszyn (1992) described the use of statistical experimental design to optimise and understand the SFE of organic contaminants from

Figure 8.3 Schematic diagram of the high-speed, thermally modulated on-line SFE–GC system (reproduced by courtesy of Professor M. Lee).

soils. They employed fractional factorial design to examine the effect of nine factors, including the extraction conditions and matrix characteristics. To study nine factors, at two levels, with eighth-order interactions, fully would require 512 experiments. However, by using their technique these authors reduced the number of experiments required to eight. Lopez-Avila and Dodhiwala (1990) used factorial design to study the extracton of PAHs from reference materials. They studied seven variables in the optimisation experiments and the results indicated that the recovery of analyte was most affected by extraction time and pressure, followed by the moisture content and size of the sample.

Bicking *et al.* (1992) employed a 2^2 factorial design to optimise extractions of EPA Method 608 pesticides from Empore™ extraction discs following preconcentration from waste water, and also for the extraction of primary amine hydrochloride pesticide from feed.

Ho and Tang (1992) employed a Simplex procedure for the optimisation of the SFE of PAHs from solid adsorbent material, and Foley (1992) compared Simplex and window diagram optimisation to SFC separations.

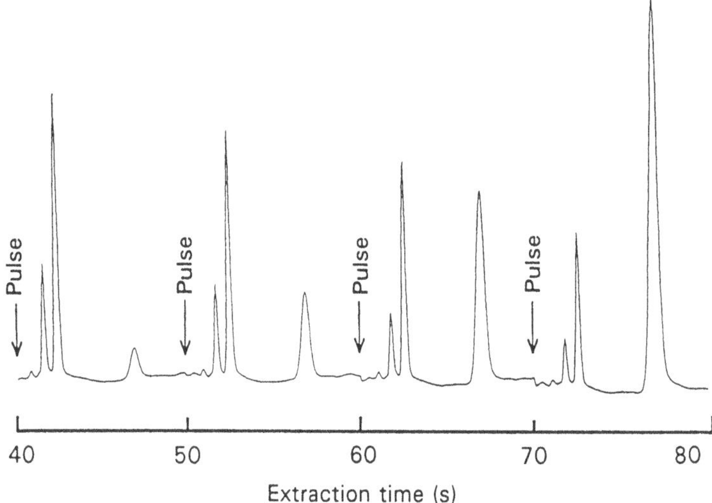

Figure 8.4 Series of four high-speed chromatograms sampled during an SFE–GC analysis of a spiked soil sample. The components in order of elution are indene, naphthalene and acenaphthene (for full experimental details see Liu *et al.*, 1992) (reproduced by courtesy of Professor M. Lee).

The application of these mathematical techniques is not yet a common occurrence, but the power of modern computers and the growing availablity of easy-to-use commercial software packages should redress this in the near future.

Comprehensive literature surveys reveal that much of the work being carried out in the development of the theory, understanding and optimisation of supercritical processes has, and still is, being carried out by process

Figure 8.5 Three-dimensional plot from an SFE–GC analysis of a spiked soil created by the summation of a series of 10 s chromatograms collected over 20 min (for full experimental details see Liu *et al.*, 1992) (reproduced by courtesy of Professor M. Lee).

and physical chemists. In order to maximise the utilisation and benefit of these developments, analysts should make efforts to develop contacts with these researchers and engage in joint research ventures to put some real science behind their experiments and remove much of the 'black-magic' tag which still haunts analytical supercritical fluid technology.

8.3 Instrumentation

In considering the possible future developments in instrumentation for analytical supercritical fluid applications it is worth briefly reviewing some of the factors that have influenced developments so far. The first major analytical application of supercritical fluids came with the introduction of SFC, which saw its main development occur during the 1980s. This development was not without conflict as many researchers appeared to be preoccupied with arguing the merits of packed versus capillary techniques, or whether a particular fluid was supercritical or not, rather than applying these methods to solve real analytical problems. However, the greatest interest in supercritical fluids has been stimulated by the emergence of SFE.

The attractions of SFE as a sample preparation technique should be apparent by now, but to date SFE instrument development has been influenced by geographical and political initiatives rather than the quality of the analytical data. This has been particularly true in the USA where the instrument manufacturers' development strategy has been largely dictated by the desire of the Environmental Protection Agency (EPA) to replace many of their traditional liquid solvent extraction methods by SFE with carbon dioxide. This initiative results from a desire to remove solvents from a health and safety point of view, rather than to achieve an improved analytical procedure. As a result instruments developed in the USA have tended to be designed with single pumps (usually syringe pumps) for use primarily with carbon dioxide. Therefore, the use of solvent-modified carbon dioxide has been limited to premixed cylinders, the addition of solvent aliquots to the pump syringe, or wetting the sample with the solvent.

In Europe and Japan, however, the main drive has been to improve the speed, selectivity and efficiency of analyte extraction, and this is reflected in instrument designs where many different pumping systems have been employed. These include reciprocating-piston high performance liquid chromatography (HPLC)-type pumps which can be readily coupled together, allowing the use of any solvent modifier at any concentration, either constant or gradient, at the touch of a button. So we can see already that the development of supercritical fluid technology to date has been steered by a number of factors, many of which have been out of the control of practising analysts who wish to use the technique. In recent times, however, there has been a noticeable shift in the development process, as more academic and industrial researchers liaise with instrument manufacturers, and we shall now consider some recent developments and trends in instrumentation.

8.3.1 Pressure control

Before any technique can be fully accepted it must be possible for analysts anywhere to carry out the same method and get the same results. Unfortunately, this has not been the case to date with SFE and to a lesser extent SFC. There are several variables that can be employed to optimise SF separations, but the density of the solvent, and therefore its pressure, is by far the most important for most applications. The effective control of pressure is therefore crucial to the success of analytical supercritical fluid techniques and it is amazing to find that the main method of achieving pressure control has been via fixed capillary restrictors constructed of fused-silica tubing (Lee and Markides, 1990). These devices do not allow independent control of flow and pressure and it is difficult to ensure reproducible results, especially with different manufacturers' instrumentation. They are also prone to blockage, particularly when they are employed for supercritical fluid extraction. If supercritical fluid techniques are to become universally applicable then a more controllable and reproducible method is required for pressure control.

The alternatives to fixed restrictors have been collectively termed variable restrictors and have come in a number of forms. These include commercially available pressure-relief valves such as those supplied by Nupro and by Rheodyne. These devices vary in size and design, but are generally set manually by varying the compression of a spring which holds the valve closed until the system pressure is sufficient to overcome the spring tension and lift the valve seat to relieve the excess pressure. These valves, however, can cause pulsation in the analytical system, which is not too much of a problem in extraction systems but is not acceptable in chromatography systems. They also tend to have too high a dead volume for use with detectors such as flame ionisation. In addition they can be prone to variations from the initial set value, due to the Joule–Thomson cooling effect from the expanding supercritical fluid. However, many workers prefer to use these instead of fixed restrictors, particularly for SFE, as they do provide some form of control over both pressure and flow.

Raynie *et al.* (1989) employed a high-pressure sheath flow nozzle where a make-up gas was added to restrict and partially control the flow from a capillary SFC column, but it is difficult to see how this could be effectively automated to provide pressure programming and flow control. More recently, Bruce *et al.* (1992) described a Variflow restrictor which is based on the radial compression of an elastic tube. It comprises an elastic tube which is compressed by an elastic ferrule in a standard high-pressure union. It is claimed that the restrictor is very tolerant of water, and the authors tested the resistance to plugging by successfully pumping 5 ml of water through it in 4 min. However, at present this is still a manually operated device and it remains to be seen whether it can be developed to be fully automated.

Saito *et al.* (1988) described an electronically controlled back-pressure regulator where a solenoid valve is employed to pulse rapidly a needle in a valve seat, thereby opening and closing the eluant flow path. In the commercial version the required back pressure is set via thumbwheels on the control module and the valve needle stroke is adjusted by a screw until the regulator operates in a smooth and continuous manner. The needle and valve assembly is surrounded by a thermostatically controlled heated collar to maintain its temperature, thus preventing blockage due to the adiabatic cooling of the expanding eluant. The rapid movement of the needle also serves to dislodge any solute particles that have been precipitated as a result of solvent depressurisation, and this has been shown to be very effective particularly for bulk extractions (Lynch, 1991), and for the collection of the products from preparative reaction studies in supercritical media (Howdle *et al.*, 1991). This regulator can also be employed for isobaric packed-column SFC, allowing the use of flow and modifier programming via the pumping system, and the use of two regulators to facilitate solvent recycling in preparative-scale SFC has also been reported (Saito and Yamauchi, 1988).

The search for total control was taken a step further with the reporting by Morrissey *et al.* (1991) of a system capable of simultaneously generating pressure and solvent-modifier gradients in packed-column SFC. They reported reproducibilities of retention times and peak areas of less than 2% relative standard deviations for several test compounds when pressure and modifier were programmed together. Superior separations for simultaneous programming over single programmes were demonstrated for polystyrene and epoxy resin samples. Pressure control was achieved via a modified self-adjusting valve equipped with an electronic control unit. System control was facilitated by a microcomputer to control the pumps and the pressure-valve control unit simultaneously.

The basic design features of this system have been developed further with a modular system comprising pumps and back-pressure control valve (Verillon *et al.*, 1991*a,b*). The system is controlled by integral microprocessors which allow simultaneous programming of pressure, mobile-phase composition and flow. This gives the analyst almost total control of the separation variables, particularly for packed-column SFC (with the exception of temperature programming). A schematic diagram of the system is shown in Figure 8.6.

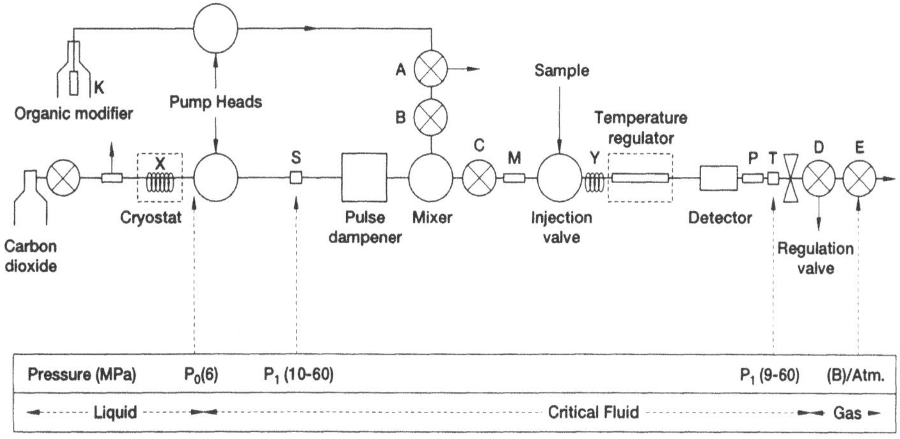

Figure 8.6 Schematic diagram of the Gilson SFC system with programmable pressure, flow and eluant composition. A to E, manual valves; K, M and P, filters; S and T, pressure transducers; X and Y, heat exchangers (reproduced by courtesy of Gilson, France).

This approach has been shown (Lynch *et al.*, 1991) to give some unique separations for a complex phenol formaldehyde resin mixture as shown in Figure 8.7. The chromatogram in Figure 8.7(a) shows the separation obtained for the mixture using a pressure programme alone with constant flow and no modifier. Figures 8.7(b) and 8.7(c) show the effects of the addition of constant modifier concentrations, 15% and 5% respectively, with the same pressure programme and flow. Figure 8.7(d), however, shows the separation obtained using simultaneous pressure, flow, and modifier-concentration programming. Comparison of these chromatograms shows that by far the best separation was achieved by using the programming of all three variables, and this simple example illustrates the advantages to be gained from effective control of the separation variables.

The commercial version of this system was launched at Pittcon '92 with claims of increased resolution and speed over HPLC, but with similar chromatographic reproducibility. Detection systems included ultraviolet (UV) and evaporative light-scattering detectors, and two example chromatograms obtained using both these detectors in series are shown in Figures 8.8 and 8.9

More recently, at the 4th International Symposium on Supercritical Fluid Chromatography and Extraction in Cincinnatti, Berger and Wilson (1992) described a similar SFC system that went one step further. This system has the capability to allow programming of pressure, density, flow, composition and temperature in packed SFC mode with a wide range of detection including flame ionisation detection (FID), UV–vis and nitrogen phosphorus detection (NPD). A schematic diagram of the system in packed mode is shown in Figure 8.10. In addition, it can be employed for

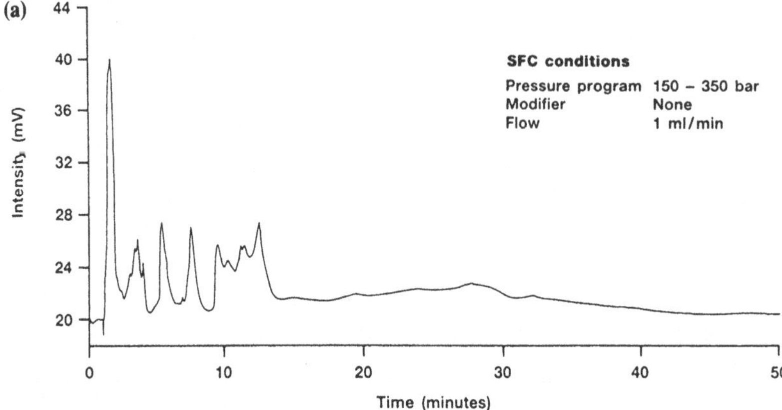

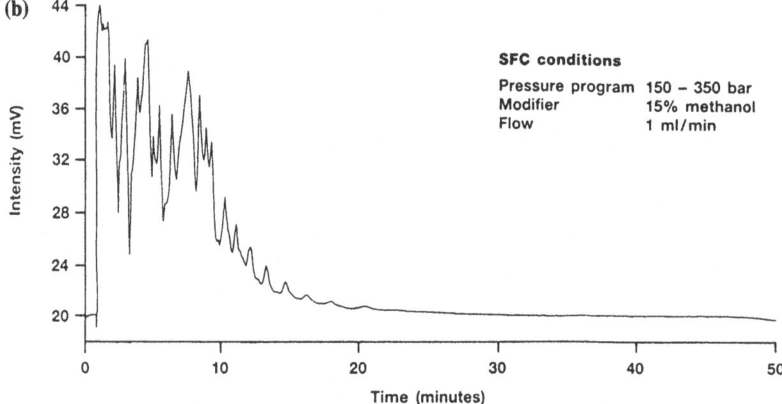

Figure 8.7(a) to (d) Series of chromatograms obtained for the SFC separation of a phenol formaldehyde resin showing the effect of pressure, modifier and flow programming. The non-variable conditions for the separation were: column: Spherisorb ODS2, 5 μm, 15 mm × 4.6 mm; temperature: 60 °C; detector: UV 280 nm, 0.5 AUFS; eluant: methanol-modified carbon dioxide (reproduced by courtesy of British Petroleum, UK).

open tubular SFC in a pressure-control mode using the configuration represented schematically in Figure 8.11.

In a second paper Berger (1992) described the use of the system for packed-column SFC. The paper described the coupling of twenty-two standard 4.6 mm × 10 cm HPLC columns packed with 5 μm particles, to produce 260 000 plates at 450 plates s^{-1} with a transit time of less than 10 min and a pressure drop of less than 160 bar. The use of the technology was then described for a number of separations including the pesticides of EPA Method 531.1, which were separated under steady-state conditions in less than 10 min and simultaneously detected using NPD and UV. The chromatograms obtained in this analysis are shown in Figure 8.12. When

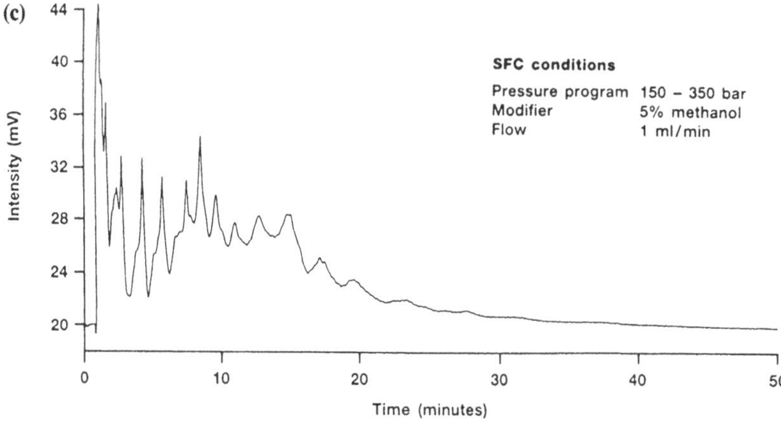

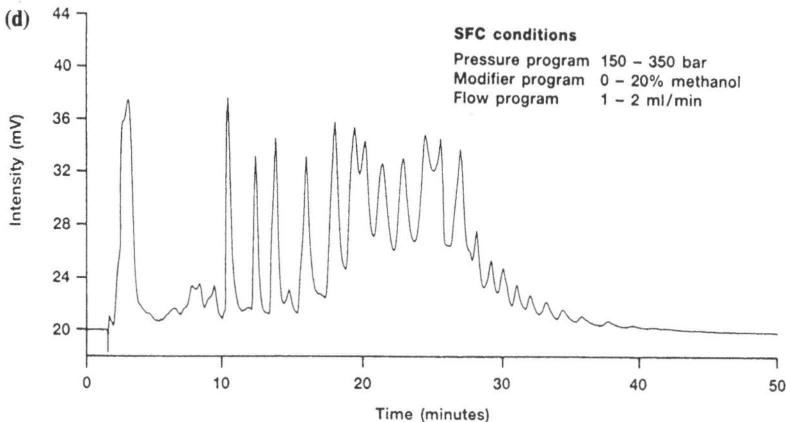

Figure 8.7 *continued*

compared to the standard LC method this gave direct selective detection five times faster with no gradient and no post-column derivatisation.

These systems are significant developments in the search for effective control of supercritical fluid operational parameters and should result in a dramatic increase in the application of SFC. They may even finally kill off the nickname of 'science fiction chromatography'! They should also stimulate research into the construction of similar devices to allow pressure and flow control both in open tubular SFC, particularly for use with FID detectors, and also in SFE with automated fraction collection.

8.3.2 Pumping systems

The essential features for pumping systems have already been covered (chapters 2 and 3), but their future development is also important for the successful application of supercritical-fluid techniques. The pumping

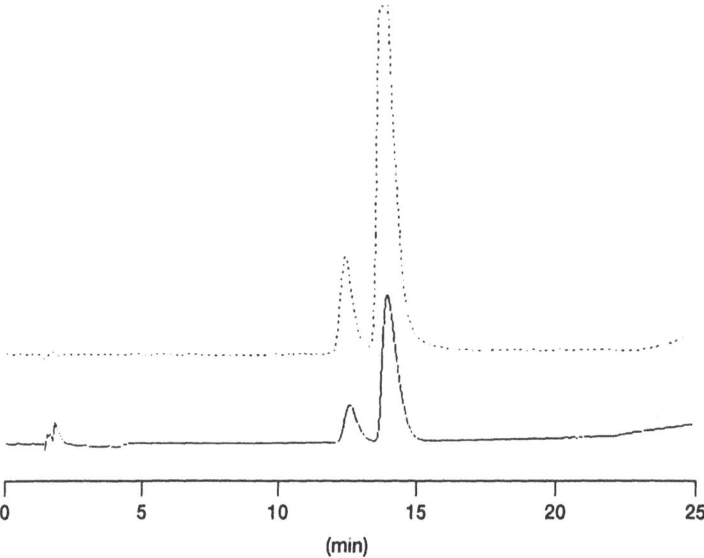

Figure 8.8 Separation of drug enantiomers by packed SFC with pressure, flow and modifier programming employing sequential UV absorbance (220 nm) (solid line) and light-scattering detection (LSD) (dashed line); column: Chiragel OG, 10 μm, 200 mm × 4.6 mm; mobile phase: carbon dioxide with 8–10% modifier in 20 min; modifier: isopropanol with 0.1% trifluoroacetic acid; total flow rate: 2–3 ml min^{-1} in 20 min; pressure (column outlet): 175–280 bar in 20 min; temperature: 45 °C (reproduced by courtesy of K. Coleman, Anachem, UK).

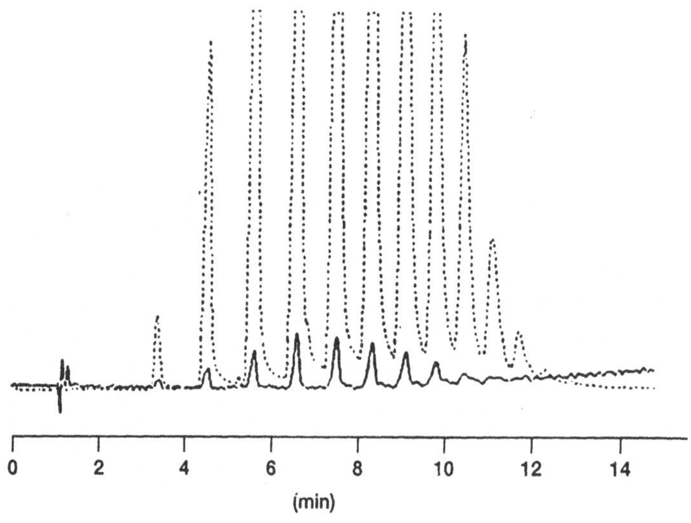

Figure 8.9 Separation of polystyrene 850 oligomers by packed SFC with pressure, flow and modifier programming employing sequential UV absorbance (245 nm) (solid line) and light-scattering detection (LSD) (dashed line); column: CN, 5 μm, 250 mm × 4.6 mm; mobile phase: carbon dioxide with 1–25% methanol in 20 min; total flow rate: 2–3 ml min^{-1} in 20 min; pressure (column inlet): 17.5–26 MPa in 20 min; temperature: 55 °C (reproduced by courtesy of K. Coleman, Anachem, UK).

Figure 8.10 Schematic representation of the Hewlett-Packard SFC instrument in packed configuration (reproduced by courtesy of Hewlett Packard).

system is the mechanism by which the fluid is delivered to the system, and development therefore must be linked closely to the development of the pressure-control devices discussed in the previous section. Ideally, we are looking for microprocessor-controlled pumps, with the control loop extending to a pressure-controlling device in the analytical system. The pumps should be able to handle a wide range of fluids and modifiers. This is particularly true for SFE, where developments in reactive extraction

Figure 8.11 Schematic representation of the Hewlett-Packard SFC instrument in open tubular configuration (reproduced by courtesy of Hewlett Packard).

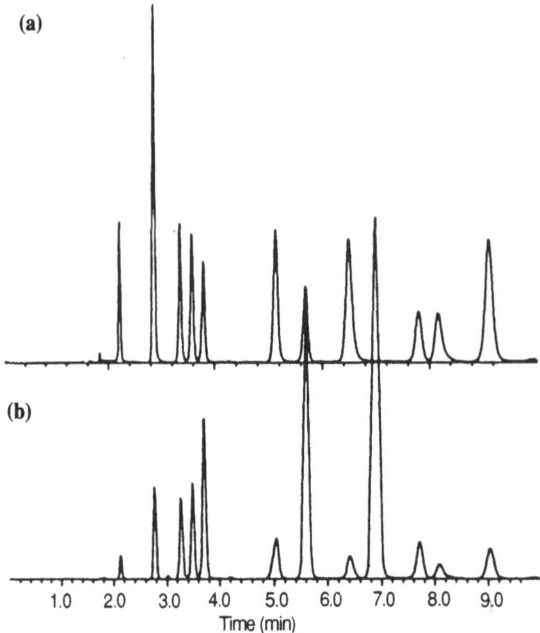

Figure 8.12 Chromatograms obtained for the separation of pesticides by EPA Method 531.1 using packed-column SFC under steady-state conditions. (a) The NPD chromatogram; (b) the UV (210 nm) chromatogram. Column: Lichrosphere Diol, 5 μm, 4.6 × 250 mm; mobile phase: CO_2 + 5% methanol; flow rate: 2.5 ml min^{-1}; temperature: 30 °C; outlet pressure: 200 bar (reproduced by courtesy of Hewlett Packard).

techniques (see section 8.4) will require pumps capable of creating fluids containing several components to facilitate efficient extraction protocols.

Syringe and reciprocating-piston HPLC-type pumps are currently used on the majority of commercial systems, with the former being more popular for capillary SFC and the latter for packed SFC particularly with modifiers. These pumps are designed mainly for use with carbon dioxide and are not generally suitable for use with other fluids, especially flammable ones. Therefore, it is likely that we will see developments in the design and construction of pumps which will address some of these deficiencies. One design that is used in larger-scale equipment and has potential as an analytical-scale pump employs a flexible diaphragm to pump the fluid.

8.3.3 Sampling systems

One of the most important factors in the successful application of an analytical technique is the requirement to be able to introduce the sample in a way that ensures the results obtained give a true representation of that sample. In addition, the continued search to achieve cost-effectiveness now requires sampling systems to be automated and capable of processing many samples without operator intervention. These criteria are particularly difficult to achieve in supercritical fluid systems due to the high pressures involved, and therefore require special considerations.

The requirements are different for chromatography and extraction, and in SFC the theory and practice of injection has been the subject of many papers and the problems are well documented elsewhere (Lee and Markides, 1990). Recent developments in large-volume solvent introduction for capillary SFC include a new solvent removal system developed by Koski *et al.* (1992), which is based on the solid-phase injectors originally developed for GC. In this system the sample solution is loaded on to a platinum wire and the solvent is evaporated. The wire is then inserted into, and sealed into, the injector where the supercritical mobile phase is introduced to solubilise the sample and introduce it on to the column.

Berg *et al.* (1992*a,b*) described the extended use of solvent venting techniques for the introduction of up to 50 μl of sample in open tubular SFC, and Cortes *et al.* (1992*a,b*) described a new system for large-volume sample injection capable of introducing up to 100 μl of liquid sample without a significant decrease in chromatographic efficiency and resolution. This system could also be employed in an in-line mode to couple LC with open tubular SFC.

In contrast, however, sample handling in SFE has not been a major area of study and the majority of work to date has involved solid samples and the design of suitable extraction cells. However, this has been mainly directed at producing cells that can be easily loaded with samples and easily coupled into the extraction system. The choice of design of extraction cells has been limited by the requirement to be able to hold pressures up to typically 500 bar safely and therefore thick-wall tubular vessels have predominated. As a result the cells tend to weigh significantly more than the sample they contain and to have a high heat capacity, but there do not seem to be any data on how long these cells require to reach thermal equilibrium and what the actual temperatures are throughout the sample during an extraction experiment. It seems inevitable that during many of the experiments to date the sample has experienced a range of temperatures during the extraction, and in some cases the centre of the cell may never have reached the selected extraction temperature. This may be particularly true when chilled pump heads are used and the fluid has not been allowed to equilibrate to the extraction temperature before being introduced to the sample cell.

Rein *et al.* (1991) reported that the chromatographic retention of PAHs spiked on a sorbent increased when using a 'long narrow vessel' as opposed to a 'short broad vessel'. They extended the work to look at the effect of microextractor cell geometry on supercritical fluid extraction recoveries of PAHs from octadecyl-bonded sorbents, and reported that extraction efficiencies were increased by more than a factor of two by decreasing the extraction cell diameter-to-length ratio from 1:20 to 1:1 (Furton and Rein, 1991).

More recently, however, Langenfield (1992) reported that they observed no significant difference in the extraction rates of native PAHs from railway-bed soil, or of flavour and fragrance compounds from lemon peel. Furthermore, they reported that neither the cell orientation (horizontal versus vertical) nor the fluid flow rate through the cell (as long as it was sufficient to sweep the cell void volume in a reasonable time) had a significant effect on extraction rates.

The whole area therefore requires further work, and it would be interesting to study sub-sampling the contents of an extraction cell from different parts, i.e. top, middle, bottom and centre out to the walls, and to determine the residual analyte concentrations after different extraction times. This would show if the hydrodynamics of the fluid flow through the cell were important.

8.4 Special techniques

In this, the final part of this chapter, we will look at some of the newer techniques that are being developed using supercritical fluids and consider the impact they may have on analytical SFE and SFC.

8.4.1 Solvents and solvent modifiers

In any literature survey on analytical supercritical fluid techniques the majority of papers will employ carbon dioxide as the main fluid with a few, mainly in packed-column SFC, describing the use of solvent modifiers. Most of the papers dealing with solvent modifiers will employ methanol as the modifier although many others have been reported (Wheeler and McNally, 1989; Berger and Deye, 1991*a,b*; Knowles and Richter, 1991).

In this section we will consider two areas that have as yet not been extensively applied in analytical supercritical fluid techniques, namely the use of fluids other than carbon dioxide and the use of 'negative modifiers'.

8.4.1.1 Alternatives to carbon dioxide. The reasons for the widespread use of carbon dioxide have been detailed previously (see chapters 2 and 3), as have its disadvantages. There are, however, many other fluids with critical temperatures and pressures within potentially usable limits (Reid *et al.*, 1987), but these have not been widely applied for a number of reasons. These include expense, toxicity, flammability, and, in the cases of ammonia and water, the very aggressive nature of the fluid in the supercritical state. However, this has not deterred many researchers who have developed techniques for handling many of these fluids and have investigated their application to analytical separations, but as yet these have not found widespread application in the industrial analytical laboratory.

Leyendecker *et al.* (1987) compared carbon dioxide, nitrous oxide, trifluoromethane, chlorotrifluoromethane, *n*-pentane, *n*-butane, isobutane, propane, ethane, diethyl ether and dimethyl ether as mobile phases for the separation of a mixture of PAHs by SFC, and Schmitz and Klesper (1990) employed different fluids for the separation of oligomers and polymers by SFC, as well as employing size-exclusion mechanisms.

Nitrous oxide has been shown to be a more efficient extraction fluid than carbon dioxide for dioxins and PAHs from fly ash and marine sediments (Alexandrou and Pawliszyn, 1989). Levy *et al.* (1991) employed nitrous oxide and sulphur hexafluoride to achieve selective extractions for the characterisation of complex environmental and petroleum matrices by coupled SFE–GC.

More recently Hawthorne *et al.* (1992*a*) reported a comparison of Freon 22 (chlorodifluoromethane), nitrous oxide and carbon dioxide for the extraction of polychlorinated biphenyls (PCBs), PAHs, and a range of non-polar to ionic analytes, and concluded that the Freon yielded consistently higher extraction efficiencies (~ 2–10 times) than either carbon dioxide or nitrous oxide. This was particularly true for samples containing high water contents.

Literature surveys reveal many more papers claiming advantages by using fluids other than carbon dioxide, but it will require further developments in instrumentation and safety procedures before these could be considered for use in a routine analytical environment. However, these other fluids should be considered as options for separations which cannot be achieved any other way.

8.4.1.2 Negative modifiers. The concept of negative modifiers or anti-solvents is an interesting one as the majority of work employing solvent modification has been directed to increasing the solubility of polar components in the supercritical fluid to facilitate more efficient extraction or elution. A negative modifier is introduced to have the opposite effect and reduce the solubility of components to achieve the desired separation. This concept was described as an aid to selectivity in SFE where nearly ideal gases were added to the fluid (Gährs, 1984), and the idea was developed further with the addition of nitrogen as a modifier for carbon dioxide in packed-column SFC (Pickel, 1991). The technique was demonstrated for the separation of aromatic hydrocarbons and proposed as a useful tool for achieving higher resolution for rapid separations with short analytical columns. It could be envisaged that a separation scheme employing a decreasing gradient of negative modifier through to pure carbon dioxide followed by a conventional solvent modifier gradient could be employed to achieve better separations for complex mixtures.

This approach could be developed further in fractionation studies where all the species of interest are dissolved in the pure supercritical fluids and are then selectively precipitated by the controlled addition of negative modifier. The application of this concept requires specialist pumping equipment that can pressurise these gases and then mix them with the supercritical fluid. As a result it will require real proof of application benefits before we can expect to see this concept in routine use.

8.4.2 *Reactions and derivatisation*

The properties of supercritical fluids that make them good solvents for extraction and chromatography can also be beneficial in performing reactions, and this has stimulated many research groups to work in this area. Much of this work is directed to producing novel compounds (Jobling *et al.*, 1990) and also to the production of high-value pharmaceuticals and chemicals for use in food, where the benign nature of carbon dioxide has obvious advantages over conventional solvents. More recently, researchers have turned their attention to applying reactions in supercritical fluids for analytical purposes. This began with the application of common derivatisation techniques used in gas and liquid chromatography (to make intractable compounds more amenable to chromatographic separation) for improving the solubilities of polar solutes in relatively non-polar solvents. This has included techniques such as methylation and the production of trimethylsilyl derivatives. These approaches were taken a step further with the increasing application of SFE where derivatisation and extraction were combined in a single process. One group (White *et al.*, 1991; Hawthorne *et al.*, 1992*b*) described the simultaneous production and extraction of fatty acid methyl esters from cellular components and their subsequent identification by GC–MS to characterise micro-organisms. The methylating agent was added to the extraction cell and the derivatisation–extraction was carried out in supercritical carbon dioxide at 400 atm and 100 °C. The products were then collected in methylene chloride and analysed by GC–MS.

Miller *et al.* (1991) reported a similar procedure to produce and extract methyl esters of phospholipids from bacteria and acid herbicides from river sediments. The benefits were claimed to include making the target analytes more amenable to extraction and also making them easier to measure. In addition, this method opens up the possibility of tagging analytes with groups to enhance their detectability, e.g. to allow the use of sensitive and selective element-specific detectors.

Hills *et al.* (1991*a,b*) performed simultaneous silylation and extraction on a sample of roasted coffee that had already been exhaustively extracted with supercritical carbon dioxide, and reported that many new components, both silylated and unsilylated, were extracted when the derivatising agent was used. The release of these underivatised components suggests that the derivatising reagent is also acting as a mechanism to release strongly-bound analytes from the matrix.

The obvious development of this type of approach is to begin to examine a wider range of complexing agents for intractable analytes. Johnston *et al.* (1992) employed experimental data and models to attempt to develop a fundamental understanding of the role of the interactions between species of interest and complexing agents that exhibit acid–base or hydrogen-bonding interactions. This type of approach should provide data that will allow analysts to select existing complexing agents or even to design new selective complexing agents to improve extraction recoveries and selectivity. One obvious application area would be the determination of polymer additives. These are generally added to polymers as sacrificial reagents to prevent degradation of the polymer, and are therefore, by their very nature, reactive species. It should therefore be relatively easy to design complexing agents that would react with these additives to form species that are more readily extractable by supercritical carbon dioxide.

8.4.3 Hyphenated techniques

The development of hyphenated techniques is an area that is showing great promise, particularly with the coupling of SFE to chromatographic techniques such as GC, LC and SFC. The advantages of coupling techniques are many but include:

- Speed of analysis
- Lowering of detection limits
- Minimising contamination
- Minimising analyte loss and decomposition
- Multicomponent data from a single shot.

These advantages have encouraged many researchers to investigate the coupling of any number of techniques. Much of the early work centred on coupling SFE with SFC, as this was technically fairly simple. However, the limited applicability of SFC led to researchers studying the coupling of SFE with GC and HPLC.

Hawthorne *et al.* (1987, 1988*a,b*, 1989), Levy *et al.* (1987), Levy and Guzowski (1988), Levy and Rosselli (1989) and others have published extensively on coupling SFE in-line with GC. Liu *et al.* (1992) have recently extended this technique with the introduction of their high-speed thermally modulated SFE–GC system described previously (section 8.2). Such coupling gains further advantages from the extensive array of detection systems that is available for GC in many laboratories. This allows the analyst to gain a large amount of data in a short time particularly from multi-hyphenated systems such as supercritical fluid extraction–gas chromatography–mass spectrometry (SFE–GC–MS), supercritical fluid extraction–gas chromatography–atomic emission detection (SFE–GC–AED), and supercritical fluid extraction–gas chromatography–Fourier transform infrared–mass spectrometry (SFE–GC–FTIR–MS).

These systems are particularly useful for the rapid examination of samples of unknown composition. For example, on-line SFE–GC–MS has been demonstrated to be a powerful tool for scanning soil samples for toxic pollutants such as PAHs, and SFE–GC–AED for the detection of trace levels of organolead residues from leaded-gasoline contamination (Lynch *et al.*, 1992).

SFE also gives GC–FTIR–MS a new lease of life in terms of increased sensitivity. The sensitivity of GC–FTIR–MS has been limited as it is generally necessary to dilute the GC column carrier gas to sweep the light pipe of the infrared. This has a knock-on effect for the mass spectrometer, as only a portion of the flow from the infrared light pipe can then be introduced to it. Thus we have a loss in sensitivity in both the detectors.

As a result the GC–FTIR–MS combination has gained a reputation as being insensitive. However, one of the benefits of hyphenated techniques is increased senstivity, and this is certainly the case with the SFE–GC–FTIR–MS where the extraction of a 10 g sample can be made equivalent to the injection of 1 μl of a solvent-based system.

The great benefit of the FTIR–MS link, however, comes with the confirmation of unknown components by both techniques. This can be particularly useful for oxygen-containing species where the strong infrared absorbances can be used to filter out non-conforming mass spectrum library searches rapidly. An example of this confirmation can be seen in Figures 8.13 and 8.14, which were obtained from the extraction of a polyethylene sample. Figure 8.13(a) shows the total ion chromatogram obtained from the mass spectrometer. Six of the peaks are readily identified as polyethylene oligomers, while the asterisked peak is the polymer additive butylated hydroxy toluene (BHT) (as shown by the sample spectrum in Figure 8.13(b) and the corresponding library match in Figure 8.13(c)). The total response chromatogram from the infrared detector is shown in Figure 8.14(a) and again six of the peaks were identified by library searches as polyethylene oligomers. The infrared spectrum of the asterisked peak is shown in Figure 8.14(b) and the best library match again identified the component as BHT. The library spectrum is shown in Figure 8.14(c) and it can be seen to be a good match. This example illustrates how the two detection systems can complement each other and combine to give confirmatory evidence for unknown component identification.

The coupling of SFE on-line with HPLC is also a powerful technique, particularly when selective detection such as mass spectrometry can be employed. SFE–HPLC–MS has been shown to give unique information for the examination of polymer additives (Lynch *et al.*, 1992) and their

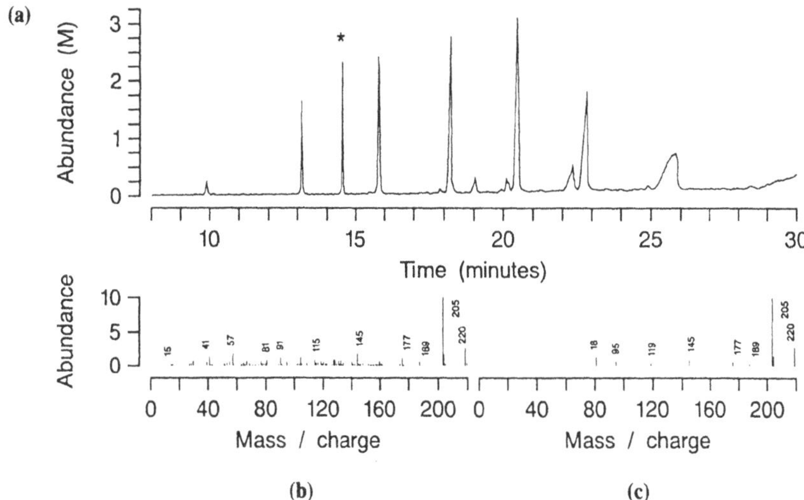

Figure 8.13 Mass spectral data from the analysis of a sample of polyethylene by SFE–GC–FTIR–MS; (a) the total ion chromatogram; (b) the mass spectrum obtained for the peak marked * in chromatogram (a); (c) the library spectrum match, which is of BHT.

breakdown products, and a schematic diagram of the system employed is shown in Figure 8.15. In this system the supercritical fluid extract is expanded into an HPLC precolumn that has been packed with the same stationary phase as the analytical column. The extracted analytes are trapped on this column and the gaseous carbon dioxide is vented. After the extraction is complete the HPLC eluant is introduced to the precolumn via

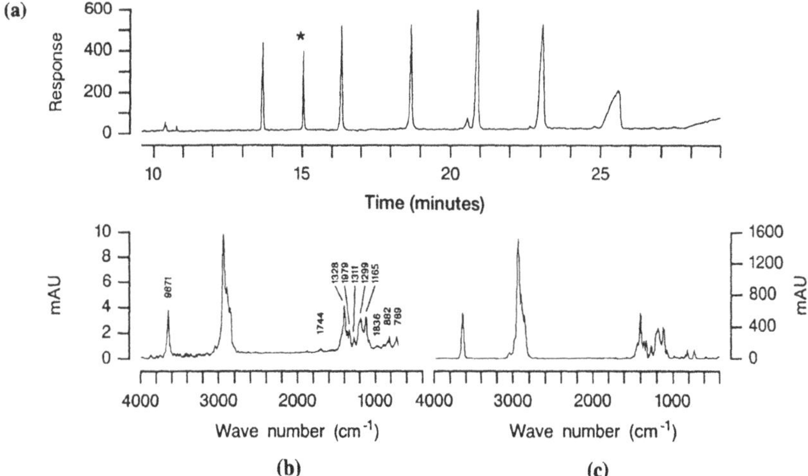

Figure 8.14 Infrared spectral data from the analysis of a sample of polyethylene by SFE–GC–FTIR–MS; (a) the total response chromatogram; (b) the infrared spectrum obtained for the peak marked * in chromatogram (a); (c) the library spectrum match, which is of BHT.

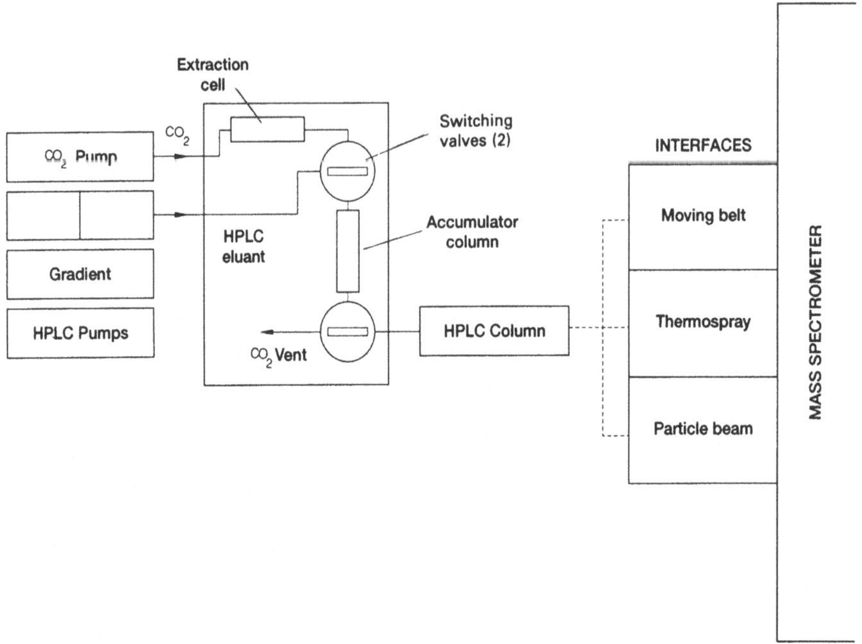

Figure 8.15 Schematic diagram of a coupled SFE–HPLC–MS system.

a valve switch and the trapped analytes are introduced into the HPLC system. The HPLC–MS interface could be operated in three modes, namely by a moving-belt interface, by a particle-beam interface, or by a thermospray interface. The particle beam–thermospray combination was particularly useful as it was available on the same instrument, and it was a simple operation to select either as required. This gave excellent flexibility as the particle beam gives electron impact (EI)-type spectra and therefore fragmentation data, and the thermospray ionisation mode provides molecular-weight data. Therefore a lot of information on unknowns could be collected in a short time. The efficiency of the SFE–HPLC–MS interfaces was assessed by the extraction of a sample of sand that had been spiked with a mixture of polymer additives; the resulting total ion chromatograms obtained by using the thermospray and particle-beam interfaces are shown in Figures 8.16(a) and 8.16(b), respectively. The mass spectra obtained from the different interfaces for Santanox are shown in Figures 8.17(a) and 8.17(b), where the different type of information obtained is clearly shown. This system has been shown to give less degradation in polymer additives than was seen when an off-line SFE with collection in solvent was employed, and this could provide a powerful tool to study the fate of reactive compounds in their natural environment.

Multi-hyphenated techniques can be extremely powerful tools for the rapid characterisation of unknown materials, and the above examples have

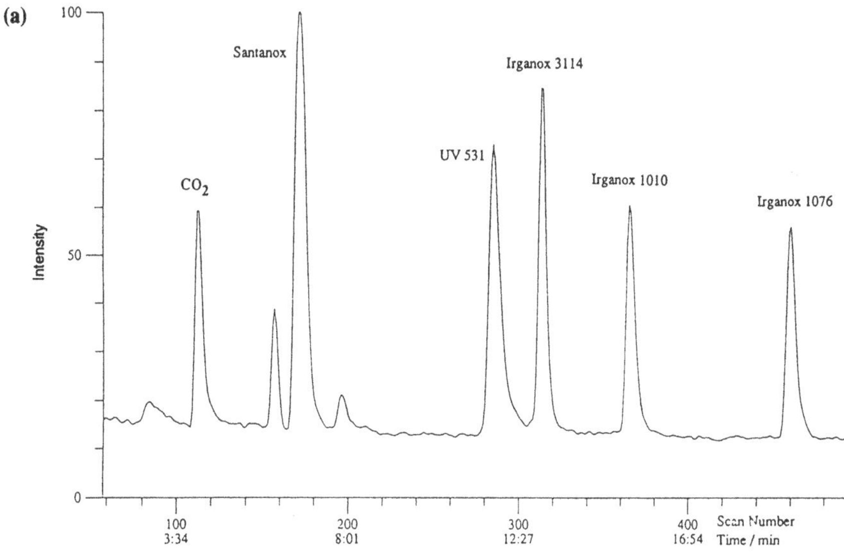

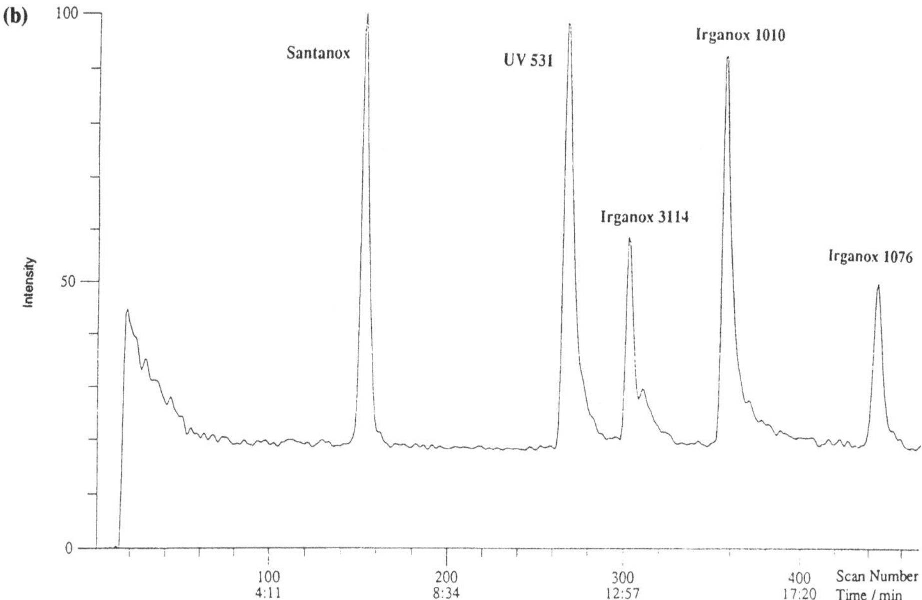

Figure 8.16 Total ion chromatograms obtained by SFE–HPLC–MS of a sample of sand spiked with a polymer additive mixture; (a) obtained using a thermospray LC–MS interface; (b) with a particle-beam LC–MS interface.

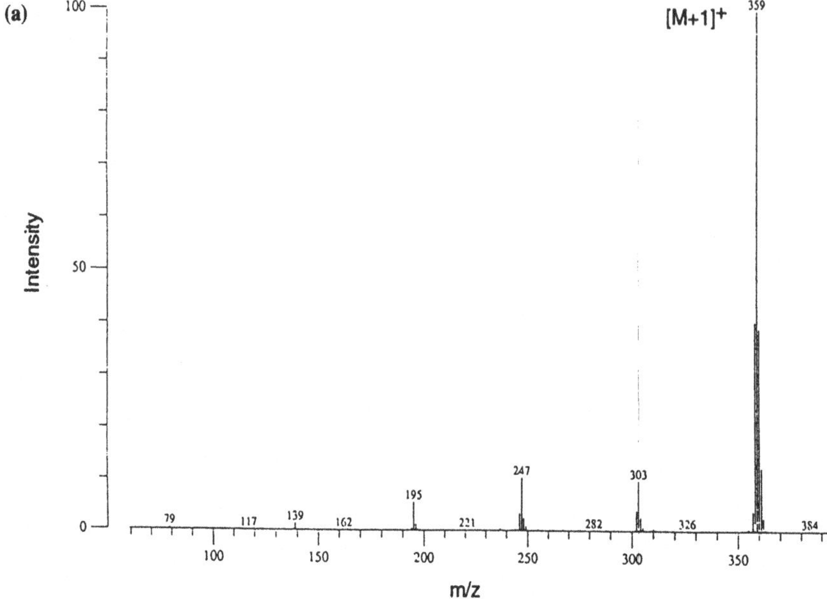

Figure 8.17 Mass spectra obtained for Santanox by SFE–HPLC–MS from a spiked sample of sand; (a) obtained using a thermospray LC–MS interface; (b) with a particle-beam LC–MS interface.

demonstrated that SFE is particularly suitable for coupling to conventional chromatographic techniques such as GC and HPLC. Furthermore, when these are combined with powerful spectroscopic detectors, unique data can be obtained and these techniques should find increasing use as routine tools in the analytical laboratory of the future.

8.4.4 Reverse micelles

One of the most exciting areas of recent research has been in the application of surfactants in supercritical fluids and in particular in the development of reverse micelles. This interest has been stimulated by the search to find ways of solubilising more polar molecules for process applications including protein extraction, biocatalysis, particle synthesis and emulsion polymerisation.

In conventional oil-in-water emulsions, the oil is the dispersed phase in the water, and forms droplets or micelles which are supported and maintained in the water phase by surfactants. In an inverse, or reverse, emulsion the water is dispersed as droplets in the oil phase, and this can provide a valuable means of introducing water-soluble components into an oil matrix. This concept has recently been extended to attempt to form reverse micelles in supercritical fluids, and much of the pioneering work in this area has been done by the groups of Smith and his co-workers at Batelle Pacific Northwest Laboratories (Gale *et al.*, 1987*a*; Fulton *et al.*, 1989; Smith *et al.*, 1990; Tingey *et al.*, 1990; Kaler *et al.*, 1991), and co-workers at the University of Texas (Johnston, 1989; Lemert *et al.*, 1990; Yazdi *et al.*, 1990). These groups have successfully demonstrated the formation of reverse micelles in supercritical hydrocarbons, and have proposed many potential applications. A supercritical pentane reverse-micelle phase with the surfactant bis-2-ethylhexylsulphosuccinate was employed as a mobile phase in packed-column SFC to separate the polar compounds phenol, 2-naphthol and resorcinol (Gale *et al.*, 1987*b*), but it is the area of extraction that offers the greatest rewards for reverse micelles in analysis. Before this can become an area of widespread study it is necessary to be able to produce reverse micelles in carbon dioxide, and this has been a major challenge.

The problem is that most of the commercially available ionic and non-ionic surfactants are ineffective in generating reverse micelles in carbon dioxide–water systems, as reported by Consani and Smith (1990), who presented observations on the solubility of over 130 surfactants in carbon dioxide. Recently, however, Hoefling *et al.* (1991) have made a significant breakthrough in this area by designing and synthesising a number of model surfactants that appear to dissolve preferentially in the carbon dioxide-rich phase of a carbon dioxide–water mixture. These surfactants were designed by using the premise that the hydrophobic tails of the surfactant intended for use in carbon dioxide should contain functional groups with low solubility parameters (silicones and fluoroethers) and low polarisability parameters (fluorinated alkanes), or which act as Lewis bases (tertiary amines), given that carbon dioxide is a weak Lewis acid. The potential of this approach was demonstrated by the extraction of the hydrophilic dye thymol blue which is completely insoluble in pure carbon dioxide at pressures up to 10 000 psi at 40 °C. This concept becomes even more exciting with the possibility of being able to design the hydrophilic head of the surfactant such that it can chemically complex the species of interest. For example, Laintz *et al.* (1991) have reported increases of several orders of magnitude in the solubilities of metal diethyldithiocarbamate complexes by substituting fluorine for hydrogen in the ligand. These exciting developments widen the potential applicability of SFE to polar components, ionic species and even heavy metals from aqueous solutions and complex matrices, and could make SFE the ultimate analytical preseparation technique.

Acknowledgements

The author would like to acknowledge the help of all his colleagues at BP who contributed to this chapter and to all the other collaborators who provided information and diagrams.

References

Alexandrou, N. and Pawliszyn, J., (1989), Supercritical fluid extraction for the rapid determination of polychlorinated dibenzo-*p*-dioxins and dibenzofurans in municipal incinerator fly ash. *Anal. Chem.*, **61**, 2770–2776.

Bartle, K.D., Clifford, A.A., Hawthorne, S.B., Langenfeld, J.J., Miller, D.J. and Robinson, R., (1990), A model for dynamic extraction using a supercritical fluid. *J. Supercrit. Fluids*, **3**, 143–149.

Berg, B.E., Flaaten, A.M., Paus, J. and Greibrokk, T., (1992*a*), Extended use of solvent venting injection techniques for large sample volumes and coupled capillary columns in SFC. *J. Microcol. Sep.*, **4**, 227–232.

Berg, B.E., Flaaten, A.M., Paus, J. and Greibrokk, T., (1992*b*), Large volume injection in capillary SFC. Basic principles and recent progress. In *Proceedings of the 4th International Symposium on Supercritical Fluid Chromatography and Extraction*, Cincinnati, OH, USA, pp. 25–26.

Berger, T.A., (1992), Recent developments in packed column SFC at Hewlett-Packard. In *Proceedings of the 4th International Symposium on Supercritical Fluid Chromatography and Extraction*, Cincinnati, OH, USA, pp. 27–28.

Berger, T.A. and Deye, J.F., (1991*a*), Separation of benzenepolycarboxylic acids by packed column supercritical fluid chromatography using methanol–carbon dioxide mixtures with very polar additives. *J. Chromatogr. Sci.*, **29**, 141–146.

Berger, T.A. and Deye, J.F., (1991*b*) Efficiency in packed column supercritical fluid chromatography using a modified mobile phase. *Chromatographia*, **31**, 529–534.

Berger, T.A. and Wilson, W.H., (1992), A new supercritical fluid chromatograph. In

Proceedings of the 4th International Symposium on Supercritical Fluid Chromatography and Extraction, Cincinnati, OH, USA, pp. 7–8.

Bicking, M.K.L., Osterheim, J.A. and Dooley, E.M., (1992), A simplified experimental design strategy for optimising SFE conditions. In *Proceedings of the 4th International Symposium on Supercritical Fluid Chromatography and Extraction*, Cincinnati, OH, USA, p. 40.

Bruce, M.L., Stephens, M.W. and Keobler, D.J., (1992), Variflow, the water proof variable restrictor. In *Proceedings of the 4th International Symposium on Supercritical Fluid Chromatography and Extraction*, Cincinnati, OH, USA, p. 41.

Bruno, T.J. and Ely, J.F. (Eds), (1991), *Supercritical Fluid Technology: Reviews in Modern Theory and Applications*. CRC Press, Boca Raton, FL, USA.

Consani, K.A. and Smith, R.D., (1990), Observations on the solubility of surfactants and related molecules in carbon dioxide at 50 °C. *J. Supercrit. Fluids*, **3**, 51–65.

Cortes, H.J., Campbell, R.M., Himes, R.P. and Pfeiffer, C.D., (1992*a*), Online coupled liquid chromatography and capillary supercritical fluid chromatography—large-volume system for capillary SFC. *J. Microcol. Sep.*, **4**, 239–244.

Cortes, H.J., Campbell, R.M., Green, L.S., Himes, R.P. and Pfeiffer, C.D., (1992*b*), On-line coupled liquid chromatography and large-volume injection system for capillary SFC. In *Proceedings of the 4th International Symposium on Supercritical Fluid Chromatography and Extraction*, Cincinnati, OH, USA, pp. 153–154.

Foley, J.P., (1992), Comparison of simplex and window diagram optimisation strategies in SFC. In *Proceedings of the 4th International Symposium on Supercritical Fluid Chromatography and Extraction*, OH, USA, pp. 97–98.

Francis, A.W., (1954), Ternary systems of liquid carbon dioxide. *J. Phys. Chem.*, **58**, 1099–1114.

Fulton, J.L., Blitz, J.P., Tingey, J.M. and Smith, R.D., (1989), Reverse micelle and microemulsion phases in supercritical xenon and ethane: light scattering and spectroscopic probe studies. *J. Phys. Chem.*, **93**, 4198–4204.

Furton, K.G. and Rein, J., (1991), Effect of microextractor cell geometry on supercritical fluid extraction recoveries and correlations with supercritical fluid chromatographic data. *Anal. Chim. Acta*, **248**, 263–270.

Gährs, H.J., (1984), Applications of atmospheric gases in high pressure extraction. *Ber. Bunsenges. Phys. Chem.*, **88**, 894–897.

Gale, R.W., Fulton, J.L. and Smith, R.D., (1987*a*), Organized molecular assemblies in the gas phase: reverse micelles and microemulsions in supercritical fluids. *J. Amer. Chem. Soc.*, **109**, 920–921.

Gale, R.W., Fulton, J.L. and Smith, R.D., (1987*b*), Reverse micelle supercritical fluid chromatography. *Anal. Chem.*, **59**, 1977–1979.

Hawthorne, S.B., (1990), Analytical-scale supercritical fluid extraction. *Anal. Chem.*, **62**, 633A–642A.

Hawthorne, S.B. and Miller, D.J., (1987), Directly coupled supercritical fluid extraction–gas chromatographic analysis of polycyclic aromatic hydrocarbons and polychlorinated biphenyls from environmental solids. *J. Chromatogr.*, **403**, 63–76.

Hawthorne, S.B., Krieger, M.S. and Miller, D.J., (1988*a*), Analysis of flavor and fragrance compounds using supercritical fluid extraction coupled with gas chromatography. *Anal. Chem.*, **60**, 472–477.

Hawthorne, S.B., Miller, D.J. and Krieger, M.S., (1988*b*), Rapid extraction and analysis of organic compounds from solid samples using coupled supercritical fluid extraction/gas chromatography. *Fresenius Z. Anal. Chem.*, **330**, 211–215.

Hawthorne, S.B., Miller, D.J. and Krieger, M.S., (1989), Coupled SFE–GC: a rapid and simple technique for extracting, identifying, and quantitating organic analytes from solids and sorbent resins. *J. Chromatogr. Sci.*, **27**, 347–354.

Hawthorne, S.B., Langenfeld, J.J., Miller, D.J. and Burford, M.D., (1992*a*), Comparison of supercritical $CHClF_2$, N_2O and CO_2 for the extraction of polychlorinated biphenyls and polycyclic aromatic hydrocarbons. *Anal. Chem.*, **64**, 1614–1622.

Hawthorne, S.B., Miller, D.J., Nivens, D.E. and White, D.C., (1992*b*), Supercritical fluid extraction of polar analytes using *in situ* chemical derivatization. *Anal. Chem.*, **64**, 405–412.

Hills, J.W., Hill, H.H. and Maeda, T., (1991*a*), Simultaneous supercritical fluid derivatisation and extraction (SFDE). *International Symposium on Supercritical Fluid Chromatography and Extraction*, Park City, Utah, USA, pp. 113–114.

Hills, J.W., Hill, H.H. and Maeda, T., (1991*b*), Simultaneous supercritical fluid derivatization and extraction. *Anal. Chem.*, **63**, 2152–2155.

Ho, J.S. and Tang, P.H., (1992), The simplex optimisation of supercritical fluid extraction of organics in a liquid–solid extraction cartridge. In *Proceedings of the 4th International Symposium on Supercritical Fluid Chromatography and Extraction*, Cincinnati, OH, USA, pp. 37–38.

Hoefling, T.A., Enick, R.M. and Beckman, E.J., (1991), Microemulsions in near-critical and supercritical CO_2. *J. Phys. Chem.*, **95**, 7127–7129.

Howdle, S.M., Jobling, M., Healy, M. and Poliakoff, M., (1991), Spectroscopic investigations of organometallic chemistry in supercritical fluids. *Symposium on Spectroscopic Investigations in Supercritical Fluids*, 201st American Chemical Society National Meeting, Atlanta, USA, Paper ANYL-2.

Huang, E.B. and Pawliszyn, J., (1992), Statistical experimental design for optimisation and understanding of SFE of organic contaminants from soil samples. In *Proceedings of the 4th International Symposium on Supercritical Fluid Chromatography and Extraction*, Cincinnati, OH, USA, p. 39.

Jobling, M., Howdle, S.M., Healy, M.A. and Poliakoff, M., (1990), Photochemical activation of carbon–hydrogen bonds in supercritical fluids: the dramatic effect of dihydrogen on the activation of ethane by $[(\eta^5\text{-}C_5ME_5)Ir(CO)_2]$. *J. Chem. Soc., Chem. Commun.*, 1287–1290.

Johnston, K.P., (1989), New directions in supercritical fluid science and technology. In *Supercritical Fluid Science and Technology*, ACS Symposium Series 406, American Chemical Society, Washington, DC, USA, pp. 1–12.

Johnston, K.P., Gupta, R., McFann, G. and Peck, D., (1992), Cosolvents, complexing agents, and organised molecular assemblies for enhanced solubilisation in SFE and SFC. In *Proceedings of the 4th International Symposium on Supercritical Fluid Chromatography and Extraction*, Cincinnati, OH, USA, pp. 33–34.

Kaler, E.W., Billman, J.F., Fulton, J.L. and Smith, R.D., (1991), A small-angle neutron scattering study of intermicellar interactions in microemulsions of AOT, water, and near-critical propane. *J. Phys. Chem.*, **95**, 458–462.

Knowles, D.E. and Richter, B.E., (1991), The use of modified mobile phases in SFC: effects on the retention and stability of selected pesticides. *J. High Resolut. Chromatogr., Chromatogr. Commun.*, **14**, 689–691.

Koski, I.J., Lee, E.D. and Lee, M.L., (1992), Solid phase injector for open tubular column SFC. In *Proceedings of the 4th International Symposium on Supercritical Fluid Chromatography and Extraction*, Cincinnati, OH, USA, pp. 13–14.

Laintz, K.E., Wai, C.M., Yonker, C.R. and Smith, R.D., (1991), Solubility of fluorinated metal diethyldithiocarbamates in supercritical carbon dioxide. *J. Supercrit. Fluids*, **4**, 194–198.

Langenfield, J.J., Burford, M.D., Hawthorne, S.B. and Miller, D.J. (1992), Effects of collection solvent parameters and extraction cell geometry on supercritical fluid extraction efficiencies. *J. Chromatogr.*, **594**, 297–307.

Lee, M.L. and Markides, K.E., (Eds), (1990), *Analytical Supercritical Fluid Chromatography and Extraction*. Chromatography Conferences, Provo, UT, USA.

Lemert, R.M., Fuller, R.A. and Johnston, K.P., (1990), Reverse micelles in supercritical fluids. 3. Amino acid solubilization in ethane and propane. *J. Phys. Chem.*, **94**, 6021–6028.

Levy, J.M. and Guzowski, J.P., (1988), Characterization of gasolines using online multidimensional supercritical fluid extraction/capillary gas chromatography. *Fresenius Z. Anal. Chem.*, **330**, 207–210.

Levy, J.M. and Rosselli, A.C., (1989), Quantitative supercritical fluid extraction coupled to capillary gas chromatography. *Chromatographia*, **28**, 613–616.

Levy, J.M., Guzowski, J.P. and Huhak, W.E., (1987), On-line multidimensional supercritical fluid chromatography/capillary gas chromatography. *J. High Resolut. Chromatogr. Chromatogr. Commun.*, **10**, 337–341.

Levy, J.M., Storozynsky, E. and Ravey, R.M., (1991), The use of alternative fluids in on-line supercritical fluid extraction–capillary gas chromatography. *J. High Resolut. Chromatogr., Chromatogr. Commun.*, **14**, 661–666.

Leyendecker, D., Leyendecker, D., Schmitz, F.P. and Klesper, E., (1987), Comparison of eluents in supercritical fluid chromatography. *J. Liquid Chromatogr.*, **10**, 1917–1947.

Liu, Z.Y., Farnsworth, P.B. and Lee, M.L., (1992), High-speed, thermally modulated SFE-GC for the analysis of volatile organic compounds in solid matrices. *J. Microcol. Sep.*, **4**, 199–208.

Lopez-Avila, V. and Dodhiwala, N.S., (1990), Supercritical fluid extraction and its application to environmental analysis. *J. Chromatogr. Sci.*, **28**, 468–476.

Lynch,T.P., (1991), Analytical applications of supercritical fluid extraction in the petroleum and petrochemical industry. *Second International Symposium on Supercritical Fluids*, Boston, MA, USA, pp. 437–440.

Lynch, T.P., Roberts, I., Escott, R.E.A. and Carrott, M.J., (1991), Analytical applications of supercritical fluid extraction in the petroleum and petrochemical industry. *Chromatographic Society Symposium on Extraction and Chromatography with Supercritical Fluids*, Keele, UK, p. 3.

Lynch, T.P., Escott, R.E.A., McDowell, P.G., Roberts, I. and Carrott, M.J., (1992), SFE for hyphenated techniques: applications from the petroleum and chemical industry. In *Proceedings of the 4th International Symposium on Supercritical Fluid Chromatography and Extraction*, Cincinnati, OH, USA, pp. 179–180.

McHugh, M.A. and Krukonis, V.J., (1986), *Supercritical Fluid Extraction: Principles and Practice*. Butterworth, Boston, MA, USA.

Miller, D.J., Hawthorne, S.B., Langenfeld, J.J. and White, D.C., (1991), SFE with chemical derivatisation for the recovery of polar and ionic analytes. *International Symposium on Supercritical Fluid Chromatography and Extraction*, Park City, UT, USA, pp. 155–156.

Morrissey, M.A., Giorgetti, A., Polasek, M., Pericles, N. and Widner, H.M., (1991), Pressure and modifier programming in packed-column supercritical fluid chromatography. *J. Chromatogr. Sci.*, **29**, 237–242.

Pickel, K.H., (1991), Influence of 'negative' modifiers by the separation of aromatic hydrocarbons. *Second International Symposium on Supercritical Fluids*, Boston, MA, USA, pp. 457–458.

Raynie, D.E., Markides, K.E., Lee, M.L. and Goates, S.R., (1989), Back-pressure regulated restrictor for flow control in capillary supercritical fluid chromatography. *Anal. Chem.*, **61**, 1178–1181.

Reid, R.C., Prausnitz, J.M. and Poling, B.E., (1987), *The Properties of Gases and Liquids*, 4th edn. McGraw-Hill, New York, NY, USA.

Rein, J., Cork, C.M. and Furton, K.G., (1991), Factors governing the analytical supercritical fluid extraction and supercritical fluid chromatographic retention of polycyclic aromatic hydrocarbons. *J. Chromatogr.*, **545**, 149–160.

Saito, M. and Yamauchi, Y., (1988), Recycle chromatography with supercritical fluid carbon dioxide as mobile phase. *J. High Resolut. Chromatogr., Chromatogr. Commun.*, **11**, 741–743.

Saito, M., Yamauchi, Y., Kashiwazaki, H. and Sugawara, M., (1988), New pressure regulating system for constant mass flow supercritical-fluid chromatography and physicochemical analysis of mass-flow reduction in pressure programming by analogous circuit model. *Chromatographia*, **25**, 801–805.

Schmitz, F.P. and Klesper, E., (1990), Separation of oligomers and polymers by supercritical fluid chromatography. *J. Supercrit. Fluids*, **3**, 29–48.

SF-Solver™, (1991), *Software for Supercritical Fluid Analysis*. Isco Inc., Lincoln, NB, USA.

Smith, R.D., Fulton, J.L., Blitz, J.P. and Tingey, J.M., (1990), Reverse micelles and microemulsions in near-critical and supercritical fluids. *J. Phys. Chem.*, **94**, 781–787.

Tingey, J.M., Fulton, J.L. and Smith, R.D., (1990), Interdroplet attractive forces in AOT water-in-oil microemulsions formed in subcritical and supercritical solvents. *J. Phys. Chem.*, **94**, 1997–2004.

Verillon, F., Heems, D., Pichon, B., Marin Martinod, T. and Robert, J.C., (1991*a*), Supercritical fluid chromatograph with independent programming of mobile phase pressure, composition and flow rate. *1st Italian Congress on Supercritical Fluids and their Applications*, Amalfi, Italy, 24–25 June.

Verillon, F., Heems, D., Pichon, B., Marin Martinod, T. and Robert, J.C., (1991*b*), Supercritical fluid chromatograph with independent programming of mobile phase pressure, composition and flow rate. *2ème Colloque sur les Fluides Supercritiques*, Paris, France, 16–17 October.

Wheeler, J.R. and McNalley, M.E., (1989), Supercritical fluid extraction and chromatography of representative agricultural products with capillary and microbore columns. *J. Chromatogr. Sci.*, **27**, 534–539.

White, D.C., Nivens, D.E., Ringelberg, D., Hedrick, D. and Hawthorne, S.B., (1991), SFE/derivatisation for rapid GC/MS identification of bacteria and bacterial products. *International Symposium on Supercritical Fluid Chromatography and Extraction*, Park City, UT, USA, pp. 43–44.

Yazdi, P., McFann, G.J., Fox, M.A. and Johnston, K.P., (1990), Reverse micelles in supercritical fluids. 2. Fluorescence and absorption spectral probes of adjustable aggregation in the two-phase region. *J. Phys. Chem.*, **94**, 7224–7232.

General index

acentric factor *see* supercritical fluids

back-pressure regulator 60, 161, 196
  see also restrictor

C_{18} extraction disks 66, 100, 132, 192
  see also solid phase extraction
capacity factor, k' 6, 19, 76, 78, 79, 80, 81, 82, 85, 89
capillary columns for SFC *see* columns for SFC, open tubular
carbon dioxide 15, 49
  adiabatic expansion 14, 59
  dip tube 12, 48
  helium over-pressure 17
  phase diagram 2, 52, 159
  pre-mixed (doped) cylinder 14, 15, 51, 160, 194
  pressure–density relationship 2, 189, 190
  properties 117, 159
  purity 15, 37, 48, 130
  quadrupole moment 159
Celite 66, 161
chemometrics
  factorial design 192
  Simplex optimization 192
chiral SFC 82, 85
cod liver oil 134
columns for SFC 12, 13, 29
  see also stationary phases
  capillary 13, 19, 23, 58, 108, 118, 195, 203
    see also open tubular
  open tubular 20, 25, 29
    film thickness 29, 109
    internal diameter 13, 29, 109
    pressure drop 108, 109
  packed columns 14, 30, 60, 107, 108, 118, 196, 197, 198
  packed capillary columns 25, 30, 36, 37, 39, 108, 133
    pressure drop 30, 108, 109, 115

derivatization *see* supercritical fluid extraction
dermatological patches 9
detection *see* supercritical fluid chromatography; supercritical fluid extraction; directly coupled
diatomaceous earth *see* supercritical fluid extraction
directly coupled
  see also hyphenated techniques
  LC–SFC 29
  SEC–SFC 133
  SFC–ICP–MS 36
  SFC–MS 39, 74, 75, 77, 87, 88, 93, 94, 95, 97, 98, 99, 108, 133
  SFC–SFC 29
  SFE–GC 61, 118, 152, 167, 169, 170, 171, 175, 179, 190, 191, 192, 193, 207, 208
  SFE–GC–FTIR–MS 61, 208
  SFE–HPLC 63, 118, 169, 173, 207, 208
  SFE–HPLC–MS 208, 210, 211, 212
  SFE–MS 63, 138
  SFE–SFC 15, 29, 61, 118, 121, 123, 124, 125, 136, 154, 169, 170, 172
  SPE–SFE 100
  SPE–SFE–GC 182

eluotropic strength 77
enzyme assay 135
equation of state 1, 6, 163
  Peng–Robinson 163
extraction methods
  see also supercritical fluid extraction
  liquid–liquid 65, 80, 117, 137
  liquid–solid 46, 65
  Soxhlet 46, 117, 122, 141, 176, 177, 182
  Soxtec 46

factorial design *see* chemometrics
fermentation broth 99
freeze drying 98, 99
  serum samples 100
fungi 99

gas chromatography–mass spectrometry 39, 132, 174, 176, 179, 184, 206

'hot-ball' model *see* supercritical fluid extraction, prediction
hyphenated techniques *see* directly coupled; multidimensional chromatography

injection *see* supercritical fluid chromatography

instrumentation for SFC *see* supercritical fluid chromatography
instrumentation for SFE *see* supercritical fluid extraction

mobile phases *see* supercritical fluids
modifiers *see* supercritical fluid chromatography; supercritical fluid extraction; supercritical fluids
molecular connectivity indices 8
multidimensional chromatography 133, 135, 136
- *see also* directly coupled
- LC–GC 135

normal phase chromatography 6, 7, 74, 85, 86, 138

phase diagram *see* carbon dioxide
pumps
- *see also* supercritical fluid chromatography; supercritical fluid extraction
- pneumatic 16
- reciprocating (piston) pump 14, 15, 17, 48, 65, 160, 194, 199, 202
  - carbon dioxide coolant 49
  - ethylene glycol coolant 49
  - pump-head cooling 17, 48, 50
- syringe 12, 13, 15, 16, 48, 49, 65, 160, 194, 199, 202

Randic algorithm 8
restrictor 12, 14, 24, 28, 30, 32, 35, 36, 47, 48, 50, 58, 60, 62, 66, 161, 172, 175, 178, 180, 182, 183, 184, 195, 196
- *see also* back-pressure regulator; supercritical fluid chromatography; supercritical fluid extraction
- ceramic frit 14, 30
- diaphragm laser drilled 59
- fixed 30, 58, 59, 195
- fused silica 30, 53, 58, 161, 180, 182
- integral 14
- linear 14, 30, 53, 59
- metal 30, 59, 161
- nozzle 60
- oscillating valve 31
  - *see also* back-pressure regulator
- pressure relief valves 195
- variable 30, 59, 67
- 'variflow' 195

Simplex optimization *see* chemometrics
solid phase extraction 66, 80, 90, 100, 132, 161, 171, 182, 184, 192
solubility in carbon dioxide
- data 5, 162, 189
- parameter 8
- parameter theory 7
- prediction 5
  - *see also* supercritical fluid extraction, prediction

solvatochromism 4
solvent polarity scale 4
solvent strength 1, 4
- *see also* solvent polarity scale

Soxhlet extraction 46
- *see also* extraction methods

stationary phases 24, 29, 62, 108, 144, 146, 171
- *see also* columns for SFC
- chiral 29, 82, 85
  - *see also* chiral SFC
- open-tubular (capillary) columns 29, 87, 108
  - film thickness 95, 96
  - polysiloxane 29, 89, 92, 95, 133, 144
- packed columns 109
- polymeric 29, 30
- porous graphitic carbon 83
- silica 30

supercritical fluids
- *see also* mobile phases
- ammonia 4, 5, 47, 160, 204
- iso-butane 205
- *n*-butane 160, 205
- carbon dioxide 2, 4, 5, 15, 16, 23, 31, 34, 36, 37, 46, 48, 49, 51, 54, 58, 64, 75, 105, 106, 117, 119, 130, 134, 143, 153, 159, 160, 204, 205, 213
  - *see also* carbon dioxide
- chlorodifluoromethane (Freon-22) 5, 160, 176, 180, 205
- chlorotrifluoromethane 160, 205
- diethylether 205
- dimethylether 205
- ethane 4, 5, 205
- methanol 4, 5
- methylene chloride 107, 120
- nitrous oxide 4, 5, 35, 51, 94, 96, 160, 175, 176, 177, 179, 180, 205
- *n*-pentane 87, 160, 205, 213
- propane 120, 143, 160, 205
- sulphur hexafluoride 35, 160
- water 5, 160, 204
- xenon 5, 33, 64

supercritical fluid chromatography 12, 106
- cryogenic focusing 62
- detectors 12, 32, 107
  - chemiluminescence 34, 114
  - electron capture 15, 37, 74, 87, 106, 107
  - flame ionization 12, 14, 15, 30, 32, 34, 35, 74, 105, 106, 107, 116, 197
  - flame photometric 36, 114
  - fluorescence 13, 14, 15, 29, 30, 32, 107

Fourier transform infrared spectroscopy 33, 106, 116, 140, 150, 151
inductively coupled plasma 36
infrared 95
ion mobility 37
laser-induced fluorescence 33
light scattering 14, 29, 32, 38, 97, 146, 197
mass spectrometry 39, 75, 106, 107, 108
microwave-induced plasma 36
nitrogen phosphorus 87, 106, 197
supersonic jet spectroscopy 33
thermionic 35, 96
UV-visible 12, 14, 15, 29, 30, 32, 75, 76, 105, 106, 107, 111, 112, 113, 116, 197
directly coupled *see* directly coupled
dynamic focusing 62
hyphenated techniques *see* directly coupled; multidimensional chromatography
ion-pair reagent 83, 84
injection 13, 17, 18, 19, 23, 28, 63, 108
direct 19, 20, 108
dynamic split injection 21, 108
split ratio 21, 114
timed-split injection 12, 21, 24, 108, 111, 112, 113, 115, 116, 126, 127, 148
instrumentation 12
method development 105, 106
peak focusing 20
retention gap 20, 24, 25, 27, 172
peak shape 23, 24, 77, 80, 83, 87, 88, 94
pressure programming 12, 197
pumps 12, 14, 16
see also pumps
solvent backflushing 26
solvent venting 24, 203
supercritical fluid extraction 46
aqueous samples *see* supercritical fluid extraction, method development
benefits 46
collection, extract 48, 59, 61, 63, 118
comparison of techniques
solid-phase extraction 100
sonication 180
Soxhlet 122, 141, 174, 176, 182
Soxhlet and sonication 174
coupled techniques
see also directly coupled; hyphenated techniques
derivatization 130, 153, 184, 206
see also supercritical fluid extraction, ion-pairing reagent
acetic anhydride 184
N-methyl-*N*-(tert-butyldimethylsilyl)-trifluoroacetamide 130
silylating agents 131, 206
tri-sil concentrate (hexamethyldisilane and trimethylchlorosilane) 131
dynamic 27, 54, 64, 67, 132, 183
extraction (sample) cell 47, 48, 52, 53, 54, 58, 61, 64, 66, 100, 118, 119, 121, 126, 160, 172, 175, 184, 203, 204
micro-cell 53, 203
packed 53
'hot-ball' model
see also supercritical fluid extraction, prediction
hyphenated techniques
see also directly coupled
in-line monitoring 54, 64, 65
instrumentation 48, 194
ion-pairing reagent
see also supercritical fluid extraction, derivatization
trimethylphenylammonium hydroxide 132
kinetics 64, 167, 175, 189
method development 6, 65, 119, 161
method development, sample type 65
method development for aqueous (liquid) samples 66, 132, 161, 182, 183, 192, 214
direct 132, 161, 182, 183
phase separator 183
method development for solid samples 132, 161, 203
freeze drying 100
grinding 66, 166
particle size 66
method development for wet samples 161, 165, 175, 179
adsorbent material 66, 171, 180, 182, 192
Celite 66, 161
diatomaceous earth 134
magnesium sulphate 161
off-line SFE 46, 58, 61, 63, 106, 117, 118, 125, 127, 136, 159, 174, 175, 176, 177, 178, 179, 180, 210
on-line SFE 46, 58, 61, 106, 117, 118, 121, 122, 123, 124, 135, 138, 159, 169, 170, 175, 179, 182, 208
operating variables 47, 66
prediction 5, 6, 55, 120, 163
chromatographic 6
equation of state 163
group contribution method (Fedors) 8
'hot-ball' model 165, 166, 167, 169, 189
molecular connectivity indices 8
molecular structure 6, 8
solubility parameter 7, 8
pumps
see also pumps

supercritical fluid extraction *cont'd*
recirculating 54
sample type *see* supercritical fluid extraction, method development
static 27, 54, 63, 64, 67, 131, 132, 135, 175, 177, 184
wet samples *see* supercritical fluid extraction, method development
supercritical fluid properties
acentric factor 189
adiabatic expansion 14, 59, 196
critical point 2, 8, 50, 58
critical pressure 1, 8, 47, 48, 105, 107, 204
critical temperature 1, 4, 19, 47, 66, 105, 107, 117, 160, 204
fractionation 46, 68, 121, 138, 153, 177
fractionation pressure 8, 120
freezing curve 1
interaction parameter
critical 7
total 7
mass transfer 3, 66, 67, 159
maximum solubility pressure 7
micelles, reverse 213
miscibility pressure 7
modifiers 5, 14, 15, 46, 49, 51, 52, 75, 76, 77, 105, 107, 120, 130, 136, 160, 194, 204
addition 5, 15, 49, 51, 66, 107, 120, 130, 137, 173
negative 204, 205, 206
phase diagram 2, 52, 159
binary fluid systems 1, 52, 159, 189
pressure–density isotherms 2, 189
pure fluids 2, 189
physical properties
diffusion 3
viscosity 3, 159
reactions 206
solubility parameter 7, 8, 67, 214
Hildebrand 189
reduced 8
solvating power 1, 114, 159
solvent polarity 4, 49, 65, 96
solvent strength 1, 4, 5, 14, 23, 47, 66, 68, 104, 105, 107, 109, 130, 142, 144
threshold pressure 66, 120, 143, 162
triple point 2

Index of compounds

applications of SFC
alkaloids 75
n-alkanes 22, 25, 26
amphetamines 76
antibiotics 98
avermectins 77
barbiturates 78
benzodiazepins 80
beta-blockers 82, 85
boron 37
cephalosporins 86
chlorine 37
cholesterol 141
cyclosporin 87
erythromycin A 87
ethoxylated alcohols 39, 111
food colours 145
carotenoids 145, 146
glycerides 25, 144
diglycerides 25, 144
monoglycerides 144
triglycerides 25, 144
hydroperoxides 141
lipids 139
mefloquine 87
mitomycin C 88
non-steroidal anti-inflammatory compounds 88
α-olefins 110
oxytetracycline 97
pesticides 36, 198
phosphorus 36
plant products 151
capsicum 152
caraway seed 152
cotton seed kernel 152
paprika 152
polycyclic aromatic hydrocarbons (PAHs) 205
chrysene 32
phenanthrene 32
polyglycols, poly(ethylene)glycol 39, 110
polymer additives 115, 121, 207
polymers 20, 106, 108, 109, 116, 205, 207
polysiloxanes 114
polystyrenes 106, 113, 196
polysulphides 114
prostglandins 91
purines 92
pyrimidines 92
steroids 94
sugars 146
sulphonamides 93
sulphur 34, 36, 37
tetracycline 97
tin, tetrabutyl 36
vegetable oils 140
vitamins 98
tocopherol 22
waxes 146
xanthines 97

applications of SFE
blood plasma 100, 132
caffeine 150, 151
coffee 150
tea 150
chlorobenzenes 174
chlorophenols 174
cholesterol 139, 141
crude oil 178
dibenzofurans 134
dioxins 134, 205
tetrachlorodibenzo-*p*-dioxins 176
2,3,7,8-TCDD 176
drug metabolite 132
essential oils 153
cold-pressed grapefruit oil 153
ethoxylated alcohols 126
see also surfactants
fish tissue 133
food colours 144
carotenoids 144
chlorophylls 145
plant material 144
fuel oil 178
glycerides 142
triglycerides 67, 134, 139, 142, 143
halocarbons 173
herbicides 132, 136
dicamba 132
(2,4-dichlorophenoxy) acetic acid 132
phenoxycarboxylic acid herbicides 178
S-triazine herbicides 177
herbs 152
basil 152
Chinese herbal medicines 152

applications of SFE *cont'd*
cinnamon 152
dodonyaro leaves 99
lemon grass 99
rosemary 152
sage 152
thyme 152
hydroperoxide 141
insecticides 135
lipids 139
mycotoxins 138
non-ionic surfactants 132
see also surfactants
organolead residues 208
pesticides 121, 134, 176, 192
plant products 99, 151
brewing hops 152
caraway seed 152
cedar wood 152
chillies 152
coffee beans, roasted 131
corn germ flour 152
cotton seed kernel 152
eucalyptus leaves 152
frankincense 152
lemon peel 152, 204
lime peel 152
Mexican spices 152
myrrh 152
nutmegs 152
orange peel 152
paprika 152
spruce needle 152
tea leaves, Japanese 131
polychlorinated biphenyls (PCBs) 121, 175, 205
polycyclic/nuclear aromatic hydrocarbons 6, 53, 121, 162, 166, 179, 190, 203, 204, 205, 208
poly(ethylene)glycols 66, 125
polymer additives 117, 119, 120, 121, 122, 208, 210
polystyrene 66, 124
steroids, megestrol acetate 8
surfactants 66, 126, 132, 213
linear alkylbenzenesulphonates 52, 166, 180
vitamins 138
waxes, hydrocarbon 123

The manufacturer's authorised representative in the EU is Springer Nature Customer Service Centre GmbH, Europaplatz 3, 69115 Heidelberg, Germany. If you have any concerns regarding our products, please contact ProductSafety@springernature.com

Printed and bound by CPI Group (UK) Ltd, Croydon, CR0 4YY
15/07/2026
02167619-0004